PROBABILITY:
AN APPROACH TO
BASIC MATHEMATICS

PROBABILITY:
AN APPROACH TO BASIC MATHEMATICS

CHUAN SENG LEE
MICHAEL HOBAN

LaGuardia Community College
CUNY

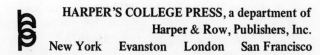
HARPER'S COLLEGE PRESS, a department of
Harper & Row, Publishers, Inc.
New York   Evanston   London   San Francisco

Library of Congress Cataloging in Publication Data

Lee, Chuan Seng.
  Probability: an approach to basic mathematics.

  1. Probabilities. I. Hoban, Michael, joint author.
II. Title.
QA273.L375   519.2   75-2078
ISBN 0-06-164401-3

Contents

Preface

This text has been developed primarily for those students who need to "brush up" their *arithmetic skills*. The approach, however, is much different from the standard "arithmetic refresher course." The essential difference is that the material uses a *problem solving approach* and practice in basic arithmetic skills arises from a need to solve practical problems.

The need to review basic arithmetic skills may arise at different times, for example, in high school, in the first year of college, or in adult-education courses. This material has been prepared to be used in any of these settings. The authors have taken great care to make the material *readable* so that students can work through it at their own pace.

The first four chapters constitute a study of elementary probability and some of its interesting applications. The bulk of the basic skills review is contained in three *Appendices*. These are located at the back of the text and are intended to be used where appropriate. This design has been adopted in order to make the use of the material as flexible as possible.

The authors believe that the needs of the students in a particular class ought to dictate how the material is used and where the emphasis is placed by the teacher. In one class, most of the time may be spent on the probability material and relatively little time on the Appendices. The student needs in another class may dictate that considerably more time be spent on the material in the Appendices and less on the probability content. The text has been prepared in this format in order to encourage this flexibility.

Presenting the bulk of the basic skills review in the Appendices has another advantage. It makes it easier to correlate this material with other aids that may be available to the students, such as programmed texts, audio or video learning aids, tutorial assistance, etc.

The material in this text has been extensively class-tested over a period of two years at LaGuardia Community College. It has been evaluated by both students and teachers and these evaluations have greatly assisted in the preparation of the present text.

A Teacher's Manual containing the answers to all exercises as well as other helpful information is available, and instructors are encouraged to consult it when using the material. The suggestions of teachers and students on how to improve the book would be greatly appreciated.

The authors wish to thank the many students, faculty members, and staff of LaGuardia Community College and Teachers College, Columbia University, who have assisted in the preparation of this text. A special word of thanks also to Caroline Eastman, of Harper & Row, Inc., for her invaluable editorial assistance in the preparation of the book.

Prologue

PROLOGUE

Joe and Bob are good friends. They often bet with each other. One day, Joe comes up with an interesting bet, which he explains as follows.

Bob will flip two pennies together. Only three things can happen: he will get two heads, two tails, or a head and a tail. Bob will win if the coins come up either two heads or two tails. Joe will win if they come up a head and a tail. Therefore, Bob has two chances to win and Joe has only one chance. So to make the bet fair, Bob has to bet two dollars against a one-dollar bet for Joe and the winner will take the three dollars. Bob agrees to the bet because he thinks that the bet is fair and that winning or losing is purely a matter of personal luck.

Would you go along with Bob's thinking? Perhaps you are more careful than Bob and would like to examine the bet situation more closely by flipping two pennies together a number of times and observing their outcomes.

Experiment: *Flipping two pennies together*

Take two pennies and flip them 20 times. Record the outcome for each flip in the following table:

Outcome	Tally	Frequency
2 heads		
a head and a tail		
2 tails		

Outcome	Tally	Frequency
2 heads	ⅢⅢ I	6
a head and a tail	ⅢⅢ ⅢⅢ	10
2 tails	IIII	4

The number of times an outcome occurs is called the *frequency* of the outcome.

Now figure out from your table the amount of money Joe and Bob would have won or lost if they had bet 20 times:

1. How many times did the outcome, a head and tail, occur out of 20?

2. How much money did Joe receive from Bob?

3. How many times did each of the two outcomes, 2 heads and 2 tails, occur?

4. How much money did Joe pay Bob?

5. How much money did Joe win from Bob, or lose to Bob?

Was Joe the winner by your calculation? If your answer is yes, do you think that it is simply because Joe is luckier than Bob?

Do the same experiment again and see who the winner is. You will probably find that Joe is the winner again. In fact, Joe is very likely to be the winner every time you flip two pennies 20 times.

Why did it happen that Joe was the winner every time you flipped two pennies 20 times?

In a *true game of chance*, do you think it is likely that Joe would be the winner *all the time*?

You are probably beginning to wonder whether the bet is unfair to Bob. How would you justify your doubt? In other words, could you point out the unfairness of the bet as stated by Joe?

Explanation of the Bet: Joe claimed that, when flipping two pennies, there are three possible outcomes: 2 heads, 2 tails, a head and a tail. It is true that these three outcomes are all that we can see on a flip of two pennies. However, did we really see what actually happened? No. *What we actually saw was confused by Joe's explanation of the bet.*

The outcome, a head and a tail, can actually occur in *two distinct ways*. That is, this outcome, simple as it looks, should be counted as *two outcomes* and *not one*. To see this point, let us take two coins of different denominations instead of two pennies.

Take a nickel and a dime, and flip them together. What can happen? You must get one of the following outcomes:

1. Two heads (a head on the nickel and a head on the dime).
2. Two tails (a tail on the nickel and a tail on the dime).
3. A head on the nickel and a tail on the dime.
4. A head on the dime and a tail on the nickel.

When flipping two coins together, we have *four possible outcomes*! Two possibilities, as listed in (3) and (4), are a head and a tail. Therefore, *Joe has two chances to win* and not one, as he claimed. Consequently, Bob and Joe should both pay the *same amount* for the bet, not two dollars to one dollar.

In games of chance, one of the most common mistakes that people make is having an incorrect idea of the total number of possible outcomes *for the game. You cannot correctly figure out your chances of winning unless you have a clear idea of how to discover the total number of possible outcomes for the game.*

Chapter I will present some techniques that will enable us to determine the total number of outcomes for a game.

Joe's explanation was not correct.

Two different ways to get a head and a tail.

Four outcomes, not just three.

Fundamental Concepts of Probability   ⌐

1.1 INTRODUCTION

In this chapter, we are going to look in a mathematical way at a few simple but intriguing games of chance such as flipping coins and rolling dice. Using these games together with our *observations* and intuition, we will develop some fundamental concepts in the modern theory of probability and statistics. These concepts will in turn be used to help us better understand the prediction of certain events such as weather forecasting and the life span of groups of people. In addition, we will gain an insight into the odds of some games of chance such as lotteries, poker, and games involving dice.

Careful observations are important in the study of probability and statistics.

In every *prediction* of an event, there always exists an element of uncertainty as to its actual outcome. It is therefore necessary that the fundamental concepts that we develop in this chapter equip us with mental tools for measuring or estimating this element of uncertainty.

We must be able to estimate the element of uncertainty.

As a first step toward this end, you will *learn how to describe your observations* in the experiments that you do in precise mathematical terms and symbols. Second, you will also *learn mathematical techniques* that will help you to list all the possible outcomes for your experiment. Third, you will *learn to assign or measure the probability of an outcome* or a set of outcomes for your experiment.

Experiment

You must be familiar with a standard deck of cards.

A *finite* number of outcomes means that the total number of outcomes is a counting number. $\{1,2,3,4,\ldots\}$

Outcome

Flipping a coin.
Two outcomes are possible: heads or tails.

Sample space

Chance Experiment: The word "experiment" is used here to mean an act, or operation, that can result in two or more outcomes, only one of which can occur at a time. Acts such as flipping a coin, rolling a pair of dice, or drawing one or more cards from a deck of 52 playing cards are examples of such experiments.

In these experiments, the outcome which does occur is usually a matter of chance. Thus, we sometimes use the term *chance experiment* to remind us of the chance or the element of uncertainty involved in our experiments.

For any experiment discussed in this book, there are only a *finite* number of possible outcomes, all of which are known before the experiment is actually done.

1.2 OUTCOMES AND SAMPLE SPACE

Before we do an experiment, we must decide what we are interested in observing in the experiment. Such *an observation* is called *an outcome* of the experiment.

For example when we flip a coin, we know from our observations that it must land on one of its sides. Thus, the outcome we would be interested in observing is *the side of the coin facing up*. Since a coin has two sides, commonly called heads and tails, the experiment of flipping a coin once has therefore only two possible outcomes: heads and tails. In other words, the experiment can result in only one of these two possible outcomes.

Definition. *A set of all possible different outcomes of an experiment is called a sample space for the experiment.*

We will describe a sample space by writing out all the possible outcomes of the experiment. The word "different" in the definition indicates that none of the outcomes we write out should be repeated.

We usually use small letters of the English alphabet or Arabic numerals to denote outcomes, and the capital letter S to denote the sample space. The total number of outcomes contained in the sample space is denoted by the symbol $n(S)$ read as "*n of S.*"

To understand the concept of sample space and to gain experience in the use of the related mathematical terms and symbols, let us consider a few games of chance. These examples will be used extensively throughout this book because they will help in understanding the mathematical concepts without unnecessary complications. Therefore, the examples should be studied very closely.

S

n (S) is a whole number

Whole number
A whole number is a number from the set $\{0,1,2,3, \dots \}$.

Four Basic Experiments:

Coin Experiment: *Flipping a coin once*

From the foregoing discussion, we may write the sample space for this experiment as:

$$S = \{h,t\}$$
Then, $n(S) = 2$

since there are only *two* possible *different* outcomes in the sample space *S*.

Note that when we write a sample space as a set of outcomes, we use the notation $\{\quad\}$, called braces, to enclose the outcomes. The outcomes enclosed within these braces are separated by a comma between them.

Examine these four experiments carefully. We will use them again.

The outcome is *t*.

{ } Braces
( ) Parentheses
[ ] Brackets

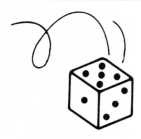

The outcome is 5.

Die Experiment: *Rolling a die once*

A die has *six* sides. Each side is marked with one through six dots. Thus, when we roll a die, one of the die's six sides must turn up. An outcome in this case is the turned-up side. The turned-up side or outcome will be designated by a numeral from 1 through 6 according to the number of dots showing. Hence we may write a sample space for this experiment as:

$$S = \{1,2,3,4,5,6\}$$
Then $n(S) = 6$,

since there are only *six* possible *different* outcomes in this sample space S.

The outcome is *t3*.

h1 means heads on the coin, 1 on the die.

Six outcomes if the die comes up *tails*.
Six outcomes if the die comes up *heads*.
t2 is *one* outcome.

Coin-die Experiment: *Throwing a coin and a die together*

If we throw a coin and a die together once, what would be one of the possible outcomes?

How many possible outcomes are there? In other words, what would a sample space for this experiment look like?

First, suppose that the coin comes up heads and the die comes up 1. Then we can write this outcome for the coin and die as *h1*. In the same way, we write *h2* for the outcome when the coin comes up heads, and the die 2. So we see that *if the coin comes up heads,* there are six possible outcomes for the coin and the die:

*h*1, *h*2, *h*3, *h*4, *h*5, and *h*6.

Of course, we know that the coin could come up tails. *If the coin comes up tails,* then we would have six *more* possible outcomes for the coin and the die:

*t*1, *t*2, *t*3, *t*4, *t*5, and *t*6.

Hence, we may write a sample space for this experiment as:

$$S = \begin{Bmatrix} h1, h2, h3, h4, h5, h6, \\ t1, t2, t3, t4, t5, t6, \end{Bmatrix}$$

Then $n(S) = 12$. Why?

Note that each outcome for this experiment consists of 2 parts. The first part shows the outcome of the coin, and the second the outcome of the die. To show that *these two parts form a single outcome*, we do not insert a comma between them. But, there must be a comma between every two outcomes enclosed in the braces.

The outcome is (3, 5).

Two-die Experiment: *Rolling a pair of dice (one green and one red)*

If we roll a green die and a red die together once, what would one of the possible outcomes be?

How many possible outcomes are there? (Without reading further, write down your guess.)

Let us suppose that the green die comes up 1 and the red die 6. If we adopt the rule of listing first the number face up on the green die and then the number on the red one, we may write this outcome for the two dice as (1,6). In the same way, we write (1,5) for the outcome when the green die comes up 1 and the red die 5. So we see that *if the green die comes up 1*, there are six possible outcomes for the two dice:

(1, 6) means 1 on the green die 6 on the red die.

We do not write this outcome as 16 because it can be mistaken for the number *sixteen*.

$$(1,1), (1,2), (1,3), (1,4), (1,5), \text{ and } (1,6).$$

Using the method just presented, *if the green die comes up 2*, we can list *another* six possible outcomes for the two dice:

$$(2,1), (2,2), (2,3), (2,4), (2,5), \text{ and } (2,6).$$

In the same way, we can list another six possible outcomes for the two dice *when the green die comes up 3*:

$$(3,1), (3,2), (3,3), (3,4), (3,5), \text{ and } (3,6).$$

Six outcomes for *each* side of the green die.

Since the *green die can come up 4,5, or 6 as well*, a sample space for this experiment may be written as:

$$S = \begin{cases} (1,1), (1,2), (1,3), (1,4), (1,5), (1,6) \\ (2,1), (2,2), (2,3), (2,4), (2,5), (2,6) \\ (3,1), (3,2), (3,3), (3,4), (3,5), (3,6) \\ (4,1), (4,2), (4,3), (4,4), (4,5), (4,6) \\ (5,1), (5,2), (5,3), (5,4), (5,5), (5,6) \\ (6,1), (6,2), (6,3), (6,4), (6,5), (6,6) \end{cases}$$

Then, $n(S) = 36$. (Was your guess correct?)

We have so far considered four experiments. Let us review here what we have done.

Every time we want to analyze an experiment mathematically, the following approach is used:

1. Identify an outcome for the experiment.

2. List all the possible different outcomes for the experiment, i.e., the sample space S.

3. Count the total number of all possible outcomes listed, i.e., $n(S)$.

From now on, this three-step approach will be followed in attempting to solve problems in this chapter.

Examples:

1. Suppose you are asked to guess a number from 0 through 9, which your friend has written down on a slip of paper. What are all the possible guesses?

Solution:

In this experiment, an outcome is a guess of a number from 0 through 9. Thus, the set of all possible different guesses, or the sample space for this guess experiment is

One outcome would be 5.

$$S = \{0, 1, 2, 3, 4, 5, 6, 7, 8, 9\}.$$

Then $n(S) = 10$. Hence the total number of possible guesses is 10.

2.   A committee of two is to be selected from a group of 5 students, 2 girls and 3 boys. How many possible committees of two are there altogether?

Solution:

In this experiment of selecting a two-student committee, an *outcome* is *a committee of two students*.

How is such an outcome denoted? Let us agree to denote the two girls as g_1 and g_2, and the three boys as b_1, b_2 and b_3. In other words, the five students are now denoted by g_1, g_2, b_1, b_2, and b_3. Thus, some examples of outcomes using these notations are $g_1 g_2$, $g_2 b_1$, and $b_2 b_3$.

An outcome is a committee of *two* students, such as $g_1 b_1$.

Note that $g_2 g_1$ would be the same committee as $g_1 g_2$.

How many such outcomes are there in all? That is, what is the sample space for this experiment?

To answer this question, let us first line up all the students as follows:

$$g_1 \; g_2 \; b_1 \; b_2 \; b_3$$

Next, write down all the committees consisting of g_1 and one other student:

$$g_1 g_2, \; g_1 b_1, \; g_1 b_2, \; g_1 b_3 \longleftarrow$$

Now, move on to g_2 to write down all the additional committees consisting of g_2:

$$g_2 b_1, \; g_2 b_2, \; g_2 b_3 \longleftarrow$$

Following this procedure, we write down all the additional committees with b_1 as a member:

$$b_1 b_2, \quad b_1 b_3$$

and the committee of $b_2 b_3$.

Hence, the sample space for this experiment is

$$S = \begin{Bmatrix} g_1 g_2, \ g_1 b_1, \ g_1 b_2, \ g_1 b_3, \ g_2 b_1, \ g_2 b_2, \ g_2 b_3, \\ b_1 b_2, \ b_1 b_3, \ b_2 b_3. \end{Bmatrix}$$

Then, $n(S) = 10$. Hence there are 10 possible different committees of two students.

EXERCISES 1.2

How well have you understood the mathematical terms and symbols?

Is the *order* of the outcomes in a sample space important?

Is it all right to repeat an outcome in the sample space?

Is the comma between two outcomes in a sample space important?

Is the *order* of the parts of an *outcome* important?

Is the order of the parts of an outcome important?

1. Explain your answer to each of the following questions:

a) Can we write the sample space $S = \{1,2,3,4,5,6\}$ for the die experiment as $\{6,4,1,3,5,2\}$?

b) Can we write the sample space $S = \{1,2,3,4,5,6\}$ for the die experiment as $\{1,2,5,6,4,3,1\}$?

c) Can we write the sample space $S = \{1,2,3,4,5,6\}$ for the die experiment as $\{123456\}$?

d) Does the symbol $n(S)$ stand for a whole number?

e) Do the symbols $(2,3)$ and $(3,2)$ stand for the same outcome for the two-die experiment?

f) Do the symbols $h6$ and $6h$ stand for the same outcome for the coin-die experiment?

g) Is passing a math course a chance experiment?

2. Write out a sample space S and find $n(s)$ for each of the following chance experiments:

   a) Drawing randomly one ball from a bag containing 4 different-colored balls (blue, green, yellow, and red). (The word "randomly" means the result of the draw is purely a matter of chance.)

   b) Drawing randomly a ball from a bag containing 2 red balls, 2 green balls, and one blue ball.

   c) Selecting randomly a flashcard from a deck of 26 flashcards, each marked with a letter from a to z.

   d) Guessing an answer to a multiple-choice question with five options (a,b,c,d,e).

   e) Guessing the month in which a stranger was born.

   f) Guessing the birthday of a stranger.

   g) Spinning the pointer of the following dial once:

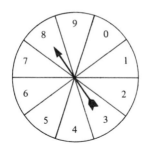

   h) Selecting randomly a coin from a bag containing a penny, a nickel, a dime, and a quarter.

   i) Selecting randomly a representative from a group of ten girls.

   j) Drawing a card from a standard deck of 52 well-shuffled playing cards.

3. Write out a sample space S and find $n(S)$ for each of the following chance experiments:

   a) Drawing randomly *two* balls (together) from a bag containing four different colored balls (blue, green, yellow, and red).

   b) Drawing randomly *two* balls (together) from a bag containing 2 red balls, 2 green balls, and one blue ball.

   c) Selecting randomly two coins (together) from a bag containing a penny, a nickel, a dime, and a quarter.

Method:
1. Identify an outcome.
2. Find all outcomes
3. Count the total number of outcomes.

Alphabet flashcards

Think of a convenient way to write down all outcomes.

We draw *two* balls at a time.
Each outcome consists of *two* parts.
Use the method from Example 2, p. 15, to write down all outcomes.

d) Flipping a nickel and a dime together once.

e) Selecting randomly *two* representatives from a group of 10 girls.

We draw *three* balls at a time. Each outcome consists of *three* parts.

4. Write out a sample space S and find $n(S)$ for the chance experiment of drawing randomly *three* balls (together) from a bag containing 3 red balls and 2 green balls.

1.3 SIMPLE AND COMPOUND EVENTS

Even	Odd
2,4,6	1,3,5

$\{0,2,4,6,8,\dots\}$ is called the set of *even* whole numbers.
$\{1,3,5,7,9,\dots\}$ is called the set of *odd* whole numbers. A whole number is either even or odd but not both.

When we perform an experiment or play a game, we are often interested in seeing the occurrence of a particular outcome. For example, in a game with a die we may win only if the number 6 occurs. Therefore, we are interested in the outcome 6.

At other times, we are interested in more than one outcome. For example, in an odd-even game with a die, we might want the outcome of a throw to be an even number. In other words, we are interested in the occurrence of the set of outcomes $\{2,4,6\}$, which is a part of the sample space $\{1,2,3,4,5,6\}$.

Occurrence of an event

The set of outcomes $\{2,4,6\}$ is said "to occur" if any one of its outcomes occurs, that is, if either 2, 4, or 6 comes up on the die.

Instead of using the phrase "a set of outcomes," we call such a set an event.

Event

Definition. Any set of outcomes *derived from the sample space for an experiment is called an* event *of the sample space.*

The capital letter S is used to denote a sample space. In the same way, we shall use capital letters to denote any event. For example, we could choose the capital letter A to stand for the event $\{2,4,6\}$.

Then we write $A = \{2,4,6\}$.

The two symbols A and $\{2,4,6\}$ are two names for the same event: the set of outcomes 2, 4, and 6. The first symbol A is convenient to use, but is not as informative as the second symbol $\{2,4,6\}$.

Two names for the same event.

The choice of the letter A is arbitrary and any other letter such as Q for $\{2,4,6\}$ might have been chosen. In that case we would have $Q = \{2,4,6\}$. However, once a letter has been chosen to denote an event, *the letter chosen must be used consistently* every time the event is referred to.

Mathematical symbols must be used *consistently.*

A closer look at the event $A = \{2,4,6\}$ and the sample space $S = \{1,2,3,4,5,6\}$ reveals that every outcome in A is also an outcome in S. This relationship between A and S is denoted by the symbol "$\subset$" and is written as "$A \subset S$" meaning

"the event A is a *subset* of the sample space S."

A

is a subset of

S

A⊂S

The concept of subset can be clarified by a more formal definition.

Definition. *Let* X *and* Y *be two events of a sample space. If every outcome in* X *is an outcome in* Y, *we say that the event* X *is a subset of the event* Y, *denoted by* X⊂Y.

Subset

In order to understand better how these terms are used, take another look at the die experiment.

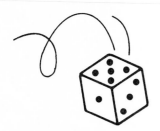

B is the event where a 2 comes up.
C is the event where a 4 *or* 5 comes up.

C

is not a subset of

A

C ⊄ A

$n(X)$ is a whole number.

Simple event *X*:
$$n(X) = 1$$

Compound event *Y*:
$$n(Y) > 1$$

Die Experiment: *Rolling a die once*

The sample space for this experiment is $S = \{1,2,3,4,5,6\}$. The event $A = \{2,4,6\}$ is the set of even numbers in the sample space.

Suppose we now have two more events $B = \{2\}$ and $C = \{4,5\}$.

Since the only outcome in B (that is, 2) is also in the event A, we say that the event B is a subset of the event A. In short, $B \subset A$.

However, such a conclusion cannot be drawn with regard to the event C. This is because there is one outcome in C, namely the outcome 5, which is not in A. Hence $C \not\subset A$, where the symbol "$\not\subset$" means "not a subset of." Note that each of the events A, B, and C is a subset of the sample space S.

Each time we deal with a particular event X, we also want to be able to discuss the total number of outcomes contained in the event X. We use the symbol $n(X)$, read as "n of X", to denote the total number.

For example, in the events $A = \{2,4,6\}$, $B = \{2\}$, $C = \{4,5\}$, we have $n(A) = 3$, $n(B) = 1$, and $n(C) = 2$, respectively.

Many events can be constructed from a given sample space. To study them, we put them into two classes, simple events and compound events.

Definition   *An event* X *is said to be* simple *if the event* X *has only one* outcome. *That is, if* n(X) = 1.

Definition   *An event* Y *is said to be* compound *if the event* Y *has* more than one *outcome. That is, if* n(Y) > 1.

(The symbol " > " means "is greater than.")

In the previous example, $n(A) = 3$ and $n(C) = 2$. Therefore, $n(A) > 1$ and $n(C) > 1$, and events A and C are compound events. However, $n(B) = 1$, so the event B is a simple event.

Greater than

Examples:

$$10 > 5$$
$$\frac{1}{2} > \frac{1}{3}$$

Example:

Two-die Experiment: *Rolling a pair of dice (one green and one red)*

(See page 14 for the sample space for this experiment.)

Suppose we have the following events:

$$D = \{(1,1)\}$$
$$E = \{(1,3), (3,1)\}$$
$$F = \{(1,1), (1,3), (2,2), (3,1)\}$$
$$G = \{(6,1), (6,2), (6,3), (6,4), (6,5), (6,6)\}$$
$$H = \{(1,1), (2,2), (3,3), (4,4), (5,5), (6,6)\}$$

Questions:

1. Which of the above events are compound events?
2. Which events are subsets of other events?

Solution:

1. All events except the event D are compound events, since each of the four events $E, F, G,$ and H has more than one outcome. To be exact, $n(E) = 2, n(F) = 4, n(G) = 6,$ and $n(H) = 6$.

2. The simple event D is a subset of the events F and H. That is, $D \subset F$ and $D \subset H$, since the outcome $(1,1)$ in D is also in F and in H.

   $D \subset F$

   $D \subset H$

   The compound event E is a subset of F. That is, $E \subset F$. Why? Is $F \subset H$? No. Why?

   $E \subset F$

Verbal description of an event. Notice that at the beginning of this section, we described the event A as $\{2,4,6\}$ and also as *the set of even numbers* from the sample space $S = \{1,2,3,4,5,6\}$. These two descriptions refer to the same event A. This is so because *one description can be translated into the other without changing the outcomes in the event.*

A *verbal description* of an event is sometimes more useful and convenient to use than the *listing description*. Thus, we need some practice in describing events in both ways.

Examples:

1. How do we describe in words the events D,E,F,G, and H in the last example?

Solution:

The event $D = \{(1,1)\}$ can be described as *the event that the sum on the two dice is two*, since the outcome $(1,1)$ can be regarded as giving a sum of two.

The two outcomes $(1,3)$ and $(3,1)$ in E seem to suggest that the event E could be described as the event that the sum on the two dice is four. However, this description would also include the outcome $(2,2)$, which is not in E. So, another way must be found to describe the event E. One way to describe the event E is as *the event that the outcome of one die is 3 and the other is 1.*

The event $F = \{(1,1), (1,3), (2,2), (3,1)\}$ could be described as *the event that the sum on the two dice is 2 or 4.*

The event $G = \{(6,1), (6,2), (6,3), (6,4), (6,5), (6,6)\}$ can be described as *the event that the outcome of the green die (the first die) is 6.* (Be sure that this verbal description gives you *precisely* the six outcomes listed in the event G.)

The event $H = \{(1,1), (2,2), (3,3), (4,4) (5,5), (6,6)\}$ could be described as *the event that the same number appears on both dice*, or *the event of getting a double.*

2. With reference to the *die-experiment*, list all the outcomes in each of the following events:

a) The event R that the outcome is an odd number.

b) The event U that the outcome is a number smaller than 5.

c) The event V that the outcome is a number larger than 6.

Solutions:

a) Since $S = \{1,2,3,4,5,6\}$ for the die experiment, and the odd numbers in S are 1,3 and 5, then the event $R = \{1,3,5\}$.

b) Since only the outcomes 1,2,3 and 4 are numbers smaller than 5, $U = \{1,2,3,4\}$.

c) It is clear from the inspection of the sample space $S = \{1,2,3,4,5,6\}$ that there is no outcome in S that would give a number larger than 6. Hence, the event V is an event which contains no outcomes, that is, $V = \{\ \}$.

In other words, V is an *empty event*. The symbol ϕ is often used to denote such an event. Thus, an empty event can be written as $\{\ \}$ or ϕ, and hence $n(\phi) = 0$.

Empty event

$$\{\ \}\ \phi$$

$$n(\phi) = 0$$

EXERCISES 1.3

1. Explain your answer to each of the following questions:

a) Does the symbol $h1$ denote an outcome or a simple event for the coin-die experiment?

b) Is $n(C)$ always greater than $n(D)$ if C is a compound event and D a simple event?

c) Does the symbol $n(A)$ stand for a whole number?

d) Can $n(A)$ be greater than $n(S)$?

e) Is $\{1,2,5\} \subset \{1,2,3,4\}$?

f) Is the set $\{7,8\}$ an event for the sample space for the die-experiment?

g) Is ϕ a simple event or a compound event?

h) If A is an event, is $\phi \subset A$?

How well have you understood the mathematical terms and symbols?
How did we define a simple event?
How did we denote a simple event?
How did we define a compound event?

$n(A) > n(S)$?

What is the sample space for the die experiment?
What does ϕ stand for?
What does $\subset$ mean?

Method:

1. Study the outcomes given to discover a rule that characterizes them.
2. Check that your rule gives you those and only those outcomes that are given.

Method:

1. Write out the sample space
2. Choose the outcomes that fit the description of the event

common outcome

subset

2. Give a description in words for each of the following events for the two-die experiment.

a) $A = \{(1,2), (2,1)\}$

b) $B = \{(1,5), (2,4), (3,3), (4,2), (5,1)\}$

c) $C = \{(1,4), (2,4), (3,4), (4,4)\}$

d) $D = \{(1,2), (2,1), (2,4), (4,2), (3,6), (6,3)\}$

e) $E = \{(1,1), (2,2), (3,3), (4,4), (5,5), (6,6)\}$

f) $F = \{(6,6)\}$

3. List all of the outcomes in each of the following events for the coin-die experiment:

a) The event A that the coin comes up heads.

b) The event B that the die comes up 1.

c) The event C that the die does *not* come up 1.

d) The event D that the die comes up 5 or 6.

e) The event E that the die comes up 3,4, or 5.

f) The event F that the coin comes up heads and the die a number less than 4.

4. In problems (a) through (e) below, let $S = \{r,b,o,y,p\}$ be a sample space for the experiment of drawing of a ball from a bag containing five different-colored balls (red, blue, orange, yellow, purple).

a) Give an event A such that $n(A) = 4$.

b) Give two compound events such that they contain only *one common outcome*.

c) Give two compound events such that they have *no common outcomes.*

d) Give two events such that one event is a subset of the other.

e) Give two events such that one event is *not* a subset of the other.

f) Give the event that contains the *largest* number of possible outcomes for this experiment.

g) Give the event that contains the *smallest* number of possible outcomes for this experiment.

5. A committee of *three* is randomly chosen from a set of two girls, denoted by g_1 and g_2, and three boys, denoted by b_1, b_2, and b_3.

   a) List a sample space of all possible committees of three.
       List the outcomes of events of this sample space for which:

   b) g_1 *and* b_1 are both on the committee.

   c) g_2 is on the committee.

   d) g_1 *or* g_2 is on the committee.

   e) b_1 *and* b_2 are both *not* on the committee.

   f) the committee has more boys than girls.

A committee has 3 members. So, each outcome has 3 parts.

The sample space may be obtained as follows:
Find all
1. committees of 3 boys.
2. committees of 2 boys and 1 girl.
3. committees of 1 boy and 2 girls.
4. committees of 3 girls.

© King Features Syndicate 1968.

1.4 PROBABILITY OF AN EVENT

Beginning with this section, you will need to make use of many arithmetic skills. There are notes in the margin to help you to remember certain facts. However, for a more complete explanation of these skills you should refer to the following Appendices:

Appendix A: Fractions, page 164
Appendix B: Decimals, page 194
Appendix C: Percents, page 212

We know that if we flip a coin, we cannot be certain that it will turn up heads. We only know that it *may* turn up heads, since heads is one of the two *possible* outcomes of the experiment. The word "may" is used to express our belief that heads is not the only possible outcome of the experiment since the coin may turn up tails as well. Now, how likely is it that the coin will turn up heads when we flip it? Is there a reason to believe that the outcome heads is *less likely*, or *equally likely*, or *more likely* to occur than tails?

In an attempt to answer these questions, let us make some observations on flipping a coin a large number of times. Suppose we flip a coin 100 times and we obtain 48 heads and 52 tails. What does this result indicate regarding our belief about the likelihood of the coin to turn up heads? Perhaps, we tend to believe that heads and tails are equally likely to occur, since the numbers of heads and tails obtained are almost equal. In other words, the result indicates strongly that there is no reason to believe that either of the two outcomes for this coin experiment is more likely to occur than the other. Once we *assume* that the two outcomes are equally likely, we say in everyday language that we have a *"50-50 chance"* or a *50% chance* of getting a head. That is, *one chance out of two* of getting a head. All three of these expressions have a striking common characteristic: They all evaluate the likelihood of getting the outcome heads *in terms of numbers*. The first expression uses the number 50 twice; the second the number 50 and 100 ($50\% = \frac{50}{100}$); and the last the numbers 1 and 2.

In fact, the last expression "one chance out of two of getting a head" can also be stated as "the chance of getting a head is $\frac{1}{2}$." The *numerator* 1 of the fraction $\frac{1}{2}$ refers to the outcome heads. To be more precise, it means that $n(A) = 1$, where $A = \{h\}$ is the event that is of interest to us.

Where did we get the *denominator* 2? The denominator 2 refers to the total number of possible outcomes for the coin. That is, $n(S) = 2$. So, we would also express the chance of getting a head as $\frac{n(A)}{n(S)}$, where $A = \{h\}$ and $S = \{h,t\}$.

Percent
50% is read as fifty percent, which means fifty out of a hundred

$$50\% = \frac{50}{100}$$

A percent is a fraction whose denominator is 100.

A complete treatment of percents is found in Appendix C.

Fraction
A fraction is a number written as $\frac{a}{b}$, where the number a is called the *numerator*, and the number b the *denominator*.

The fraction $\frac{a}{b}$ also means $a \div b$

Let us see whether we can apply our analysis for the coin-experiment to the die experiment.

Recalling the die experiment, suppose we bet on the outcome 5. What chance do we have to get the outcome 5 in one throw of a die?

There are six possible outcomes of which 5 is one. We say that "my chance is one out of six of getting a 5." This numerical expression also means the same as "my chance of getting a 5 is $\frac{1}{6}$."

Where did we get the numbers 1 and 6? Since the outcome 5 is the only outcome that we bet on, this means that we are interested in the occurrence of the event $B = \{5\}$. Thus, the numerator 1 refers to $n(B) = 1$. The number 6 refers to $n(S) = 6$, since there are only six possible outcomes for a die. We would therefore express our chance of getting the outcome 5 as $\frac{n(B)}{n(S)}$, where $B = \{5\}$ and $S = \{1,2,3,4,5,6\}$.

Our chance then of getting a 5 when we roll a die can be represented by the fraction $\frac{1}{6}$. We also know that *a fraction can be changed into a percent*. In this case

$$\tfrac{1}{6} \approx 17\%$$

(We read this as $\frac{1}{6}$ *"is approximately equal to"* 17%.)

In expressing the chances of getting the outcome 5, we have already assumed one important fact: *all the six possible outcomes are equally likely to occur*. In other words, there is no reason for us to believe that one outcome is more or less likely to occur than any of the other five. *This assumption is crucial in our way of calculating the chances of getting the outcome 5.*

Why is it that we believe each of the six outcomes for the die is equally likely to occur? A practical way to see this is to throw a die a large number of times and observe the *frequency* of each outcome in relation to the total number of throws. Suppose we decide to throw a die 360 times.

Proper fraction
$(b > a)$
$\frac{3}{7}, \frac{2}{9}, \frac{5}{27}$, etc.

Improper fraction
$(a > b)$
$\frac{7}{3}, \frac{9}{2}, \frac{27}{5}$, etc.

Mixed Fraction
$2\frac{1}{3}, 4\frac{1}{2}, 5\frac{2}{5}$, etc.

A complete treatment of fractions is found in Appendix A.

See Sec. C2, into a percent for changing a fraction

Approximately equal

Equally - likely assumption

Frequency

$$\frac{360}{6} = 60 \text{ times}$$

How many times would you expect each of the six outcomes to occur out of 360 throws?

If the six outcomes are all equally likely to occur, then we would expect that each of the six outcomes would occur approximately the same number of times, that is, about 60 times apiece.

Once we are convinced by our observations that the six outcomes are equally likely to occur, we would then use the fraction $\frac{n(X)}{n(S)}$ for calculating the chance that an event would occur.

Definition. *Suppose that all outcomes of a sample space* S *for an experiment are* equally likely to occur. *If* X *is any event of* S, *then the chance* of the event X occurring *is equal to the fraction* $\frac{n(X)}{n(S)}$.

Probability of an event X:

$$P(X) = \frac{n(X)}{n(S)}$$

In notation, we write the definition as

$$P(X) = \frac{n(X)}{n(S)}$$

The symbol $P(X)$, read as "P of X," means the *chance* of the event X occurring; to be more mathematical, it means *the probability of the event X occurring.* From now on, we will use the word "probability" for "chance." The word "probability" has the advantage of not having unnecessary meanings that we might have for the word "chance." So, if $A = \{h\}$ for the coin experiment, then we would say that "the probability of the event A is $\frac{1}{2}$" instead of "the chance of the event A occurring is $\frac{1}{2}$."

$A = \{h\}$

$P(A) = \frac{n(A)}{n(S)} = \frac{1}{2}$

$B = \{5\}$

$P(B) = \frac{n(B)}{n(S)} = \frac{1}{6}$

Similarly, if $B = \{5\}$ for the die experiment, we would say that "the probability of the event B is $\frac{1}{6}$" instead of "the chance of the event B occurring is $\frac{1}{6}$."

Examples:

1. Die Experiment: *Rolling a die once*

a) What is the probability of the event $C = \{3\}$ occurring? $n(C) = 1$

b) What is the probability of the event $D = \{5,6\}$ occurring? $n(D) = 2$

c) What is the probability of the event $E = \{1,3,5\}$ occurring? $n(E) = 3$

Solution:

Since $S = \{1,2,3,4,5,6\}$ and $n(S) = 6$, then

a) $P(C) = \dfrac{n(C)}{n(S)}$

$= \dfrac{1}{6}$, since $n(C) = 1$

$\approx 17\%$

b) $P(D) = \dfrac{n(D)}{n(S)}$

$= \dfrac{2}{6}$

$= \dfrac{1}{3}$

$\approx 33\%$

c) $P(E) = \dfrac{n(E)}{n(S)}$

$= \dfrac{3}{6}$

$= \dfrac{1}{2}$

$= 50\%$

$\dfrac{1}{6} = 1 \div 6$

$$\begin{array}{r} .166 \approx .17 = 17\% \\ 6\overline{)1.000} \\ \underline{6} \\ 40 \\ \underline{36} \\ 40 \\ \underline{36} \\ 4 \end{array}$$

See Sec. A.4 for reducing a fraction.

See Sec. B.4 for rounding off a decimal.

See Sec. C.2 for changing a fraction to a percent.

Note that the preceding three examples show that *the more outcomes an event has, the higher is its probability of occurring.* Thus, $P(E) > P(D)$, and $P(D) > P(C)$.

2. Coin-die Experiment: *Throwing a coin and a die together*

a)   What is the probability of a simple event A occurring?

b)   What is the probability of the event that the coin comes up heads?

$\dfrac{1}{12} = 1 \div 12$

$\begin{array}{r} .083 \\ 12\overline{)1.000} \\ \underline{96} \\ 40 \\ \underline{36} \\ 4 \end{array}$

$\dfrac{1}{12} \approx .083$

$\approx .08,$ why?

$= \dfrac{8}{100}$

$= 8\%$

Note: $.08 \neq .8,$ since

$.08 = \dfrac{8}{100}$

$.8 = \dfrac{80}{100}$

See Sec. B.2

Solution:

Recall that for this experiment:

$S = \{h1,\ h2,\ h3,\ h4,\ h5,\ h6,\ t1,\ t2,\ t3,\ t4,\ t5,\ t6\}$ and $n(S) = 12$.

Since all the 12 outcomes are equally likely to occur and $n(A) = 1$, we therefore have:

a) $P(A) = \dfrac{n(A)}{n(S)}$

$= \dfrac{1}{12},$ Why?

$\approx 8\%$

b)   We have discussed in exercise 3 of Exercises 1.3 that the event that the coin comes up heads is the same as the event $\{h1,\ h2,\ h3,\ h4,\ h5,\ h6\}$.
Let $B = \{h1,\ h2,\ h3,\ h4,\ h5,\ h6\}$. Then

$P(B) = \dfrac{n(B)}{n(S)} = \dfrac{6}{12}.$ Why?

$= \dfrac{1}{2}$

$= 50\%,$ which means that we should get one of the outcomes contained in the event B about half of the time. For example, in 100 throws of a coin and a die, we will get approximately 50 outcomes that are contained in B.

3. Two-die Experiment: *Rolling a pair of dice (one green and one red)*

   a)   What is the probability of a simple event A occurring?

   b)   What is the probability of the event that the sum on the two dice is 5?

   c)   What is the probability of the event that the same number comes up on both dice?

Solution:

Since $n(S) = 36$ (see Section 1.2), and all the 36 outcomes are equally likely to occur, we therefore have

   a) $P(A) = \dfrac{n(A)}{n(S)}$

             $= \dfrac{1}{36}$, Why?

             $\approx 3\%$

   b)   The event that the sum on the two dice is 5 is the same as the event $B = \{(1,4),\ (2,3),\ (3,2),\ (4,1)\}$. Thus,

$$P(B) = \frac{n(B)}{n(S)}$$

$$= \frac{4}{36}, \text{Why?}$$

$$= \frac{1}{9}$$

$$\approx 11\%$$

   c)   The event that the same number comes up on both dice is the same as the event

$$C = \{(1,1),\ (2,2),\ (3,3),\ (4,4),\ (5,5),\ (6,6)\}.$$
        Thus,

$$P(C) = \frac{n(C)}{n(S)}$$

$$= \frac{6}{36}, \text{Why?}$$

$$= \frac{1}{6}$$

$$\approx 17\%$$

$\dfrac{1}{36} = 1 \div 36$

$$36\overline{)1.000} ^{.027}$$

$\dfrac{1}{36} \approx .027$

$\approx .03$, Why?

$= \dfrac{3}{100}$

$= 3\%$

See Sec. C.2

$\dfrac{17}{100} > \dfrac{11}{100}$

See Sec. A.3 for Comparison of
Fractions.

Of the two events B and C, we see that the event C is more likely to occur than the event B, since $P(C) > P(B)$, that is $17\% > 11\%$.

The committee must have 2 students on it.

So, an outcome consists of 2 students.

4. **Selecting a committee of two.** *A committee of two is to be selected at random from a group of 5 students, 2 girls and 3 boys.*

a) What is the probability that a committee of 2 girls will be chosen?

b) What is the probability that a committee of 1 girl and 1 boy will be chosen?

c) What is the probability that a committee of 2 boys will be chosen?

Solution:

See pp.15-16.

$g_1 \, g_2$ is a committee with 2 girls on it.

Let us denote the two girls as g_1 and g_2, the 3 boys as b_1, b_2, and b_3. Then

$$S = \{g_1g_2, \ g_1b_1, \ g_1b_2, \ g_1b_3, \ g_2b_1, \ g_2b_2, \\ g_2b_3, \ b_1b_2, \ b_1b_3, \ b_2b_3\} \text{ and } n(S) = 10$$

Are the 10 outcomes all equally likely to occur? Yes, since a committee is chosen at random. Then we can use our definition for probability.

A is a simple event.
$n(A) = 1$

$\dfrac{1}{10} = \dfrac{1 \times 10}{10 \times 10}$

$\quad = \dfrac{10}{100}$

$\quad = 10\%$

See Sec. C.1.

a) Since the outcome in S that means a committee of 2 girls is g_1g_2, we therefore have

$$A = \{g_1g_2\} \text{ and, } P(A) = \frac{n(A)}{n(S)}$$

$$= \frac{1}{10}$$

$$= 10\%$$

b) Since the outcomes in S that mean a committee of 1 girl and 1 boy are g_1b_1, g_1b_2, g_1b_3, g_2b_1, g_2b_2, and g_2b_3, we therefore have

$$B = \{g_1b_1,\ g_1b_2,\ g_1b_3,\ g_2b_1,\ g_2b_2,\ g_2b_3\} \text{ and}$$
$$P(B) = \frac{n(B)}{n(S)}$$
$$= \frac{6}{10}$$
$$= 60\%$$

$n(B) = 6$

c) Since the outcomes in S that mean a committee of 2 boys are b_1b_2, b_1b_3 *and* b_2b_3, we therefore have

$$C = \{b_1b_2,\ b_1b_3,\ b_2b_3\} \text{ and}$$
$$P(C) = \frac{3}{10}$$
$$= 30\%$$

$n(C) = 3$

Therefore, it is *most* likely to choose a committee of 1 girl and 1 boy.

$$60\% > 30\% > 10\%$$
$$\text{or } \frac{60}{100} > \frac{30}{100} > \frac{10}{100}$$

See Sec. A.3.

How well have you understood the mathematical terms and symbols?

Examine the fraction $\dfrac{n(X)}{n(S)}$ more closely.

$X = S$

$X = \phi$

How large can $\dfrac{n(X)}{n(S)}$ be?

How small can $\dfrac{n(X)}{n(S)}$ be?

Method:
1. Find S and $n(S)$.
2. Find X and $n(X)$.
3. Find $P(X)$.
4. Express answer in percent.

Method:
1. Find S and $n(S)$.
2. List the outcomes in the given event X.
3. Find $n(X)$.
4. Find $P(X)$.
5. Express answer in percent.

Use the method suggested in Exercise 3.

EXERCISES 1.4

1. Explain your answer to each of the following questions:

   a) Can 0.1 be the probability of some event?

   b) Can 0.001 be the probability of some event?

   c) Is $P(S) = 1$?

   d) Is $P(\phi) = 0$?

   e) Can $\dfrac{17}{16}$ be the probability of some event?

   f) Is $P(A)$ a fraction between 0 and 1 inclusive?

2. In the die experiment, what is the probability *in percent* of:

   a) A given simple event X?

   b) The event $A = \{1,2,\}$?

   c) The event $B = \{1,2,3,4,5\}$?

3. In the coin-die experiment, what is the probability *in percent* of:

   a) A given simple event X?

   b) The event that the die comes up 1?

   c) The event that the coin comes up heads?

4. A ball is drawn randomly from a bag containing 5 different-colored balls: black, blue, green, red, and white. What is the probability *in percent* of each of the following events:

   a) $\{$ red $\}$

   b) $\{$ blue $\}$

   c) $\{$ black, red $\}$

   d) $\{$ black, red, white $\}$

   e) $\{$ any color except blue $\}$

5. In the two-die experiment, what is the probability *in percent* of the event that the *sum* on the two dice is:

a) 2   e) 6   i) 10

b) 3   f) 7   j) 11

c) 4   g) 8   k) 12

d) 5   h) 9

Method:

1. Recall S and n (S) for the two-die experiment
2. List all outcomes of the event X described by the numeral
3. Find $n(X)$.
4. Find $P(X)$.
5. Express answer in percent

6. In Exercise 5,

a) Which event is most likely to occur?

b) Which are the events that are least likely to occur?

c) Which events are equally likely to occur?

Comparison of fractions denoting probabilities.

7. A card is drawn *at random* from a deck of 52 well-shuffled playing cards. What is the probability *in percent* that you will draw:

a) The ace of hearts

b) An ace

c) A spade

d) A red card

e) A picture card (Jack, Queen, King)

f) A number card (A,2,3,4,5,6,7,8,9,10)

g) A card whose number is less than 6

h) A card which is not an ace

i) A card which is either a 4 or a 5 of any suit

The phrase "at random" means the result of a draw is purely a matter of chance.

Use the method suggested in Exercise 5.

Name: _____

Experiment I: *Flipping a coin 30 times*

AIM: *To show that the two outcomes, heads and tails, are equally likely to occur.*

A. First Experiment:

Flip a coin 30 times and record the results.

If the outcome heads occurs 20 *times out of 30 flips*, then the *relative frequency* of the outcome heads is $\frac{20}{30} = \frac{2}{3}$.

In this case, f_1 = 20. Note that the *relative frequency* is expressed as a fraction.

Outcome	Tally	Frequency (f_1)	Relative Frequency $(\frac{f_1}{30})$
Head			
Tail			
Sum			

1. Which of the two outcomes occurred more often?

_____

2. What is the difference between the two relative frequencies?

_____

B. Second experiment:

Your first experimental results very likely showed that one side of the coin occurred more often than the other. Check out these results by actually flipping the coin another 30 times.

Outcome	Tally	Frequency (f_2)	Relative Frequency $(\frac{f_2}{30})$
Head			
Tail			
Sum			

3. Which of the two outcomes occurred more often?

   _____

4. What is the difference between the two relative frequencies?

   _____

5. Is the answer that you have just given to question 3 the same as the answer given to question 1?

   _____

C. Using the results in these two experiments that you have just performed, complete the table below:

Outcome	Frequency $(f_1 + f_2)$	Relative Frequency $(\dfrac{f_1 + f_2}{60})$
Head		
Tail		
Sum		

6. What is the difference between the two relative frequencies?

   _____

7. Which of the three answers given in questions 2, 4, and 6 is the smallest?

   _____

8. What answer to question 7 would you expect if the two outcomes, heads and tails, are equally likely to occur?

   _____

CONCLUSION: *Have you achieved the aim of the experiment? Explain.*

_____

_____

_____

_____

Name: ───────────────────────

Experiment II: *Throwing a die 30 times*

AIM: *To show that the six outcomes, 1,2,3,4,5 and 6, are equally likely to occur.*

A. First Experiment:

   Throw a die 30 times and record the results.

Outcome	Tally	Frequency (f_1)	Relative Frequency $(\frac{f_1}{30})$
1			
2			
3			
4			
5			
6			
Sum			

1. Which of the six outcomes occurred most often?

   ─────────────────────────────────────

2. Which of the six outcomes occurred least often?

   ─────────────────────────────────────

3. Which of the six relative frequencies is the largest?

   ─────────────────────────────────────

4. Which of the six relative frequencies is the smallest?

   ─────────────────────────────────────

5. What is the difference between the largest and the smallest frequencies?

   ─────────────────────────────────────

B. Second Experiment:

If you throw the same die another 30 times, would you predict that the outcome you have given in question 1 will occur most often, and that the outcome in question 2 will occur least often?

_____

Test your predictions.

Outcome	Tally	Frequency (f_2)	Relative Frequency $(\frac{f_2}{30})$
1			
2			
3			
4			
5			
6			
Sum			

6. Which of the six outcomes occurred most often?

_____

7. Which of the six outcomes occurred least often?

_____

8. Are your predictions correct?

_____

9. Which of the six relative frequencies is the largest?

_____

10. Which of the six relative frequencies is the smallest?

_____

11. What is the difference between the largest and the smallest relative frequencies?

_____

C. Using the results in these two experiments that you have just performed, complete the table below:

Outcome	Frequency $(f_1 + f_2)$	Relative Frequency $(\dfrac{f_1 + f_2}{60})$
1		
2		
3		
4		
5		
6		
Sum		

12. Which of the six relative frequencies is the largest?

_____

13. Which of the six relative frequencies is the smallest?

_____

14. What is the difference between the largest and the smallest relative frequencies?

_____

15. Which of the three differences given in questions 5, 11, and 14 is the smallest?

_____

16. What would you expect the answer to question 15 to be, if the six outcomes, 1,2,3,4,5, and 6, are equally likely to occur?

_____

CONCLUSION: *Have you achieved the aim of the experiment? Explain.*

_____

_____

_____

_____

By permission of John Hart and Field Enterprises Inc.

1.5 EQUALLY LIKELY OUTCOMES

In exercise 5 of Exercises 1.4 we saw in throwing two dice once, that the sum on the two dice ranges from the smallest sum 2 to the largest sum 12. This means that the set $\{2,3,4,5,6,7,8,9,10,11,12\}$ represents a set of all possible outcomes (sums) when two dice are thrown once. Hence it is a sample space for the two-die experiment (see the definition for a sample space given in Section 1.2). But, are all the outcomes in this sample space $S_1 = \{2,3,4,5,6,7,8,9,10,11,12\}$ equally likely to occur? The answer is no.

From exercise 5 of Exercises 1.4, we have calculated that:

$$P(\{2\}) = \frac{1}{36} \qquad P(\{8\}) = \frac{5}{36}$$

$$P(\{3\}) = \frac{2}{36} \qquad P(\{9\}) = \frac{4}{36}$$

$$P(\{4\}) = \frac{3}{36} \qquad P(\{10\}) = \frac{3}{36}$$

$$P(\{5\}) = \frac{4}{36} \qquad P(\{11\}) = \frac{2}{36}$$

$$P(\{6\}) = \frac{5}{36} \qquad P(\{12\}) = \frac{1}{36}$$

$$P(\{7\}) = \frac{6}{36}$$

These results show that the sum of 7 is most likely to occur and the sum of 2 or 12 is least likely to occur. Hence we conclude that the 11 outcomes are not equally likely to occur. Note that,

Sum on two dice:
$S_1 = \{2,3,4,5,6,7,8,9,10,11,12\}$

Equally-likely outcomes?

7 is *most likely*.

2 or 12 are *least likely*.

with the exception of the outcomes 2 and 12, each of the other outcomes   3,4,5,6,7,8,9,10,   and   11   in   the   sample   space $S_1 = \{2,3,4,5,6,7,8,9,10,11,12\}$ is a *compound event* of another sample space:

$$S = \begin{cases} (1,1), & (1,2), & (1,3), & (1,4), & (1,5), & (1,6) \\ (2,1), & (2,2), & (2,3), & (2,4), & (2,5), & (2,6) \\ (3,1), & (3,2), & (3,3), & (3,4) & (3,5), & (3,6) \\ (4,1), & (4,2), & (4,3), & (4,4), & (4,5), & (4,6) \\ (5,1), & (5,2), & (5,3), & (5,4), & (5,5), & (5,6) \\ (6,1), & (6,2), & (6,3), & (6,4), & (6,5), & (6,6) \end{cases}$$

We will show this by actual observations.

In Exercises 1.5 these results will be verified by throwing two dice 60 times. You will soon discover that you will get the sum of 7 most often, and the sum of 2 or 12 least often.

Observe that the sample space $S_1 = \{2,3,4,5,6,7,8,9,10,11,12\}$ violates the basic assumption of the definition for the probability of an event (see pg. 28) because all outcomes of the sample space must be equally likely to occur. Therefore the definition does not apply to a sample space such as this. The sample space to be used in calculating the probability of an event must be the *primitive sample space.*

Primitive sample space

Definition. *A sample space for an experiment is called a* primitive sample space *if all possible outcomes in the sample space are equally likely to occur.*

Probability of a simple event

A simple event has only 1 outcome.

According to this definition, the probability of a simple event of a primitive sample space is always equal to $\frac{1}{n(S)}$, since the number of outcomes in a simple event is equal to 1.

Examples:

1. The probability of any simple event in the coin experiment is $\frac{1}{2}$, since the primitive sample space $S = \{h,t\}$ has only two possible outcomes, that is, $n(S) = 2$.

2. The probability of any simple event in the die experiment is $\frac{1}{6}$ since the primitive sample space $S = \{1,2,3,4,5,6\}$ has only six possible outcomes, that is, $n(S) = 6$.

3. The probability of any simple event in the coin-die experiment is $\frac{1}{12}$. Why?

4. The probability of any simple event in the two-die experiment is $\frac{1}{36}$. Why?

Let us now consider the experiment in the Prologue where two coins were tossed together. Let $X = \{0,1,2\}$ where "0" means "no heads," "1" means "one head," and "2" means "two heads."

Is the set $X = \{0,1,2\}$ a sample space for this two-coin experiment?

The answer is yes, because when two coins are tossed, the result is either no heads, 1 head, or 2 heads. Hence, a set of all possible outcomes for the two-coin experiment is $X = \{0,1,2\}$.

Is the sample space $X = \{0,1,2\}$ a primitive sample space for the two-coin experiment? That is, are all the three outcomes in X equally likely to occur?

X is a sample space but *not a primitive* sample space.

The answer is *no*. We will perform an experiment at the end of this section by throwing a penny and a nickel 30 times to verify that the outcome "1 head" in $X = \{0,1,2\}$ will occur most often.

We will show this by actual observations.

Let us now re-state the definition for the probability of an event in terms of the primitive sample space:

Definition. *Suppose that* S *is a* primitive sample space *for an experiment. If* X *is an event from* S, *then the* probability of the event X *occurring is*

Probability of an event

$$P(X) = \frac{n(X)}{n(S)}$$

Example:

A ball is randomly drawn from a bag containing 6 balls: 1 blue, 2 green, and 3 red balls.

    a)   What is the probability that a blue ball will be drawn?

    b)   What is the probability that a green ball will be drawn?

    c)   What is the probability that a red ball will be drawn?

Solution:

To determine the probability of each event, we must first write out a *primitive* sample space.

Before we can answer these questions, let us first figure out the *primitive sample space* for this experiment of drawing one ball randomly.

There are only three types of colored balls: blue, green, and red. So, a sample space for this experiment is $\{b, g, r\}$. But, these three outcomes are not equally likely to occur, since, for example, there are more red balls than green or blue. Thus, we need a sample space that will take account of the number of balls in each color. It can be assumed that the two green balls are distinguishable in some way, and so are the three red balls.

Let g_1 and g_2 denote the two green balls, and r_1, r_2 and r_3 the three red balls. Now, we have

primitive sample space

$$S = \{b,\ g_1,\ g_2,\ r_1,\ r_2,\ r_3\}$$

In this case, each of the six outcomes in S is equally likely to occur (or to be drawn). Therefore, we have a primitive sample space S for this experiment.

With this primitive sample space, we can now use the probability formula and answer the three questions above.

A is the event that the blue ball is drawn.

a)

$$\text{Let } A = \{b\}$$
$$\text{Then } P(A) = \frac{n(A)}{n(S)}$$
$$= \frac{1}{6}$$

b)
$$\text{Let } B = \{g_1, g_2\}$$
since a green ball means the outcome g_1 or g_2.

$$\text{Then } P(B) = \frac{n(B)}{n(S)}$$
$$= \frac{2}{6}$$
$$= \frac{1}{3}$$

B is the event that one of the two green balls is drawn.

c)
$$\text{Let } C = \{r_1, r_2, r_3\}. \text{ Why?}$$
$$\text{Then } P(C) = \frac{n(C)}{n(S)}$$
$$= \frac{3}{6}$$
$$= \frac{1}{2}$$

C is the event that a red ball is drawn.

These three results agree with our intuitive feelings that a red ball has the highest probability of being drawn, a green ball the next highest probability, while the blue ball is least likely to be drawn.

Since $\frac{3}{6} > \frac{2}{6} > \frac{1}{6}$,

then $\frac{1}{2} > \frac{1}{3} > \frac{1}{6}$

See Sec. A. 3.

EXERCISES 1.5

1. A ball is drawn at random from a bag containing 2 white, 3 red, 4 green, and 5 blue balls. Find the probability that the ball drawn is:

   a) white
   b) red
   c) green
   d) blue
   e) white *or* red
   f) green *or* blue
   g) not green
   h) neither red nor blue
   i) white or green or blue

Method:
1. Devise a way to describe the outcomes so that they constitute the primitive sample space S.
2. Find all outcomes in the given event X, and find $n(X)$.
3. Find $P(X)$

2. A box contains five balls, each marked with a numeral from 1 to 5. *Two* balls are drawn together at random and their numbers are *added*. Find the probability of the event that the *sum* of the two numbers is:

Two balls are drawn at a time.
Each outcome has two parts.
Use the method suggested in Exercise 1.

a) 3   c) 5   e) 7   g) 9
b) 4   d) 6   f) 8

Three letters are selected.
Each outcome has 3 parts. Write out the primitive sample space.

3. *Three* letters are selected at random from the word "money." Find the probability of the event that:

   a)   One of the letters selected is "n"

   b)   Two of the letters selected are vowels.

How many students on a committee? What is the primitive sample space?

4. A committee of *three* has to be chosen by drawing lots in a class consisting of 3 girls and 3 boys. Find the probability that the committee will consist of:

   a)   3 girls

   b)   2 girls and 1 boy

   c)   1 girl and 2 boys

   d)   *at least* one girl

   e)   not more than 2 boys

   f)   3 boys

1 card is drawn.

5. A card is drawn randomly from a standard deck of 52 well-shuffled playing cards.

   Find the probability of each of the events given below:

   a) $A = \{$an ace$\}$

   b) $B = \{$a spade$\}$

   c) $C = \{$a picture card$\}$

   d) $D = \{$a red card$\}$

   e) $E = \{$not a heart$\}$

All whole numbers that have the number 2 as a *factor* are called *multiples of 2.*

Examples:

2 = 2 X 1
4 = 2 X 2
6 = 2 X 3
8 = 2 X 4
8 : a multiple of 2
2 : a factor of 8

6. In the two-die experiment (Section 1.2), list all the outcomes in each of the following events for which the *sum* on the two dice is:

   a)   A multiple of 2

   b)   A multiple of 3

   c)   A multiple of 5

   d)   A multiple of 6

   e)   A multiple of 2 or 3

7. Find the probability of each of the events (a-e) defined in Exercise 6.

Name: ————————————————————

Experiment III: *Flipping a penny and a nickel together 30 times*

AIM: *To show that the sample space $\{0,1,2\}$ is NOT a primitive sample space for this experiment.*

A. Let us first consider the sample space $S = \{hh,ht,th,tt\}$, where the first letter of the symbol for each outcome in S denotes the outcome of the penny, and the second letter the outcome of the nickel. Flip the two coins together 30 times and record your results in the table.

Outcome	Tally	Frequency (f)	Relative Frequency ($\frac{f}{30}$)	R.F. in Percent
hh				
ht				
th				
tt				
Sum				

1. Are the four relative frequencies equal?

————————————————————————————

B. Now let us consider the sample space $S = \{0,1,2\}$, where 0, 1, and 2 mean no heads, 1 head, and 2 heads respectively. Using the results in your experiment, complete the table below:

Outcome	Frequency (f)	Relative Frequency ($\frac{f}{30}$)	Relative Frequency in Percent
2			
1			
0			
Sum			

2.   Which of the three relative frequencies is the largest?

     _____

3.   Is each of the three outcomes equally likely to occur?

     _____

CONCLUSION: *Have you achieved the aim of the experiment?*
              Explain.

     _____

     _____

     _____

     _____

     _____

Name: _____

Experiment IV: *Rolling a pair of dice (one green and one red) together 50 times*

AIM: *To show that S = {2,3,4,5,6,7,8,9,10,11,12} is NOT a primitive sample for this experiment.*

A. Consider the sample space S with $n(S) = 36$ as shown in the table (the notation (1,5) means "1 on the green die and 5 on the red die"). Roll the dice 50 times and complete the table.

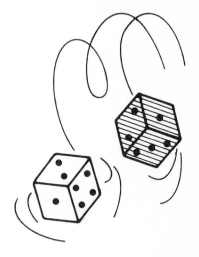

Outcome	Tally	Frequency	Outcome	Tally	Frequency
(1,1)			(4,1)		
(1,2)			(4,2)		
(1,3)			(4,3)		
(1,4)			(4,4)		
(1,5)			(4,5)		
(1,6)			(4,6)		
(2,1)			(5,1)		
(2,2)			(5,2)		
(2,3)			(5,3)		
(2,4)			(5,4)		
(2,5)			(5,5)		
(2,6)			(5,6)		
(3,1)			(6,1)		
(3,2)			(6,2)		
(3,3)			(6,3)		
(3,4)			(6,4)		
(3,5)			(6,5)		
(3,6)			(6,6)		
Sum			Sum		

B. Next, use the experimental results which you have just obtained to write down the frequency of each outcome in another sample space {2,3,4,5,6,7,8,9,10,11,12}, where the numeral 7, for example, means that the *sum* of the two numbers on the two dice is 7.

Outcome	Frequency (f)	Relative Frequency ($\frac{f}{50}$)	Relative Frequency in Percent
2			
3			
4			
5			
6			
7			
8			
9			
10			
11			
12			
Sum			

1. Which of the 11 outcomes has the largest corresponding relative frequency?

   _____

2. Which of the 11 outcomes has the smallest corresponding relative frequency?

   _____

3. Give a pair of outcomes that have the same corresponding relative frequency.

   _____

4. Is each of the 11 outcomes in this sample space equally likely to occur?

   _____

CONCLUSION: *Have you achieved the aim of the experiment? Explain.*

_____

_____

_____

_____

_____

ANOTHER LOOK AT THE PROLOGUE

Now that we have examined some of the basic ideas of probability, we can take a better look at the bet that Joe proposed to Bob in the Prologue to Chapter 1.

Primitive sample space for the two-coin experiment.

We now know that a *primitive* sample space for the two-coin experiment would be written as $S = \{hh, ht, th, tt\}$. Bob wins if either of the two outcomes, two heads or two tails, comes up. So the event favorable to Bob is $B = \{hh, tt\}$. And

$$P(B) \ = \ \frac{2}{4} \ = \ \frac{1}{2}$$

Joe wins if the outcome, a head and a tail, comes up. So the event favorable to Joe is $J = \{ht, th\}$. And

$$P(J) \ = \ \frac{2}{4} \ = \ \frac{1}{2}$$

Same probability.

From this it can be seen that *both Bob and Joe have the same probability of winning.* Therefore, the bet is fair only if they each wager the same amount of money and the winner takes all. This is not, of course, the way Joe explained it.

Expected Cost of a game. See Chap. 3

But what happens if two people make a bet where the probabilities are not equal? Is there any way of making such a bet fair to each person? The answer is yes. The question deals with the concept of *expected cost of a game*—which will be examined in the third chapter.

SUMMARY

Experiments:

1. *Coin Experiment: Flipping a coin once*

$$S = \{h,t\}$$
$$n(S) = 2$$

2. *Die Experiment: Rolling a die once*

$$S = \{1,2,3,4,5,6\}$$
$$n(S) = 6$$

3. *Coin-die Experiment: Throwing a coin and a die together*

$$S = \{h1,\ h2,\ h3,\ h4,\ h5,\ h6,\ t1,\ t2,\ t3,\ t4,\ t5,\ t6\}$$
$$n(S) = 12$$

4. *Two-coin Experiment: Flipping two coins together (a penny and a nickel)*

$$S = \{hh, ht,\ th,\ tt\}$$
$$n(S) = 4$$

5. *Two-die Experiment: Rolling a pair of dice together (one green and one red)*

$$S = \begin{Bmatrix} (1,1),\ (1,2),\ (1,3),\ (1,4),\ (1,5),\ (1,6) \\ (2,1),\ (2,2),\ (2,3),\ (2,4),\ (2,5),\ (2,6) \\ (3,1),\ (3,2),\ (3,3),\ (3,4),\ (3,5),\ (3,6) \\ (4,1),\ (4,2),\ (4,3),\ (4,4),\ (4,5),\ (4,6) \\ (5,1),\ (5,2),\ (5,3),\ (5,4),\ (5,5),\ (5,6) \\ (6,1),\ (6,2),\ (6,3),\ (6,4),\ (6,5),\ (6,6) \end{Bmatrix}$$
$$n(S) = 36$$

Terms and Symbols:

Terms	Symbols	Page
Chance experiment		10
Outcome		10
Equally likely outcomes		41
Sample Space	S	10
Primitive sample space	S	42
Number of outcomes in a (primitive) sample space	$n(S)$	42
An event	Any capital letter except S	18
Simple event A	$n(A) = 1$	20
Compound event B	$n(B) > 1$	20
Empty event	$\emptyset, \{ \ \}$	23
An event X is a subset of the event Y	$X \subset Y$	19
An event A is *not* a subset of the event B	$A \not\subset B$	20
Probability of an event X	$P(X)$	28
Randomly (at random)		17, 35
Frequency of an outcome	f	4
Relative frequency of an outcome		36
Approximately equal	$\approx$	27
Greater than	$>$	21
Braces	$\{ \ldots \}$	11

Formula:

Let S be a primitive sample space for an experiment.

If $X \subset S$, then

$$P(X) = \frac{n(X)}{n(S)}$$

6. *The probability that a particular sum will occur when we roll a pair of dice.*

$$n(\{2\}) = n(\{12\}) = 1 \qquad P(\{2\}) = P(\{12\}) = \frac{1}{36}$$
$$n(\{3\}) = n(\{11\}) = 2 \qquad P(\{3\}) = P(\{11\}) = \frac{2}{36}$$
$$n(\{4\}) = n(\{10\}) = 3 \qquad P(\{4\}) = P(\{10\}) = \frac{3}{36}$$
$$n(\{5\}) = n(\{9\}) = 4 \qquad P(\{5\}) = P(\{9\}) = \frac{4}{36}$$
$$n(\{6\}) = n(\{8\}) = 5 \qquad P(\{6\}) = P(\{8\}) = \frac{5}{36}$$
$$n(\{7\}) = 6 \qquad\qquad P(\{7\}) = \frac{6}{36}$$

Properties of Probability:

1. $P(\phi) = 0$ (0%)

2. $P(S) = 1$ (100%)

3. $P(X)$ is a fraction between 0 and 1 inclusive, given that $X \subset S$. $P(X)$ is a percent between 0% and 100% inclusive, given that $X \subset S$.

REVIEW EXERCISES

1. State *one main difference* and give one example for each pair of the mathematical terms:
   a) Event; Sample space.
   b) Simple event; Compound event.
   c) Sample space; Primitive sample space.
   d) Event; Probability of an event.

2. A letter is chosen at random from the word "sampling." Which of the following sets is the *primitive* sample space for the experiment and which are just sample spaces?
   a) { vowel, consonant }
   b) { vowel, s, m, p, l, n, g }
   c) { s, a, m, p, l, i, n, g }
   d) { s, a, m, l, i, n, g }

3. List all possible events of the sample space $S = \{a, b, c\}$.

4. A letter is chosen at random from the word "random."

   a) What is a primitive sample space for this experiment?

   b) Give the event which consists of all the vowels in the word "random."

   c) What is the probability of the letter chosen being a vowel?

   d) Give the event which consists of all the consonants in the word "random."

   e) What is the probability of the letter chosen being a consonant?

   f) Is a consonant more likely to be chosen than a vowel? Why?

5. Two players simultaneously show their right hands to each other, exhibiting one or two or three extended fingers. If we assume that each player is equally likely to extend one, two, or three fingers,

   a) What is the primitive sample space for this experiment?

   b) What is the probability that the total number of fingers extended is even?

   c) What is the probability that the total number of fingers extended is less than 4?

6. A committee of *two* (a chairman and a secretary) is to be chosen from a group of five students: Carol, Gladys, Jack, Bob, and Paul.

   a) How many different committees of two can be chosen?

   b) What is the probability that a committee of two will consist of Carol and Bob?

   c) What is the probability that a committee of two will consist of Carol and Gladys?

   d) What is the probability that a committee of two will have Carol as one of the two students?

7. There are six balls in a bag each having one of the following letters on it: k, l, m, n, o, p. You are asked to draw one ball from the bag at random.

   a) What is the primitive sample space for this experiment?

   b) Write out an event Q such that $n(Q) = 3$

   c) Write out an event T such that T is a subset of Q.

   d) What is the probability of the event Q?

   e) What is the probability of the event T?

   f) Which of the fractions in (d) and (e) is larger?
      How much larger?

8. With reference to the *two-die experiment* which has the primitive sample space where $n(S) = 36$,

   a) Find the probability of *not* throwing a double.

   b) Find the probability that the number on one die is double the number on the other.

   c) Find the probability that one die gives a 5 and the other die a number less than 5.

   d) Which is the smallest probability of the three probabilities given in (a), (b), and (c)?

9. An American roulette wheel has 38 equally-spaced compartments numbered from 00, 0, 1,2,3, ... , 35, 36 around its rim. The two compartments numbered 00 and 0 are colored green. Eighteen of the remaining 36 compartments are colored red and the rest are colored black. While the wheel is spinning, a small ball is dropped on it. Assuming that the ball is equally likely to rest on any one of the 38 compartments,

   a) What is the probability *in percent* that the ball will rest on the compartment numbered 00 or 0?

   b) What is the probability *in percent* that the ball will rest on a red compartment?

   c) What is the probability *in percent* that the ball will rest on a non-red compartment?

   d) What is the probability *in percent* that the ball will rest on one of the compartments numbered 1 through 15?

10. Assume that boys and girls have an equal probability of being born. In a family of two children,

   a) What is the probability that both children are boys?

   b) What is the probability that both children are girls?

   c) What is the probability that one is a boy and the other is a girl?

   d) Which of the three probabilities is the largest?

   e) Is this problem mathematically the same as the two-coin experiment? Explain.

Name: ————————————————————————

Experiment V: *Throwing a coin and a die together 30 times*

AIM: *To show that each of the 12 outcomes is equally likely to occur.*

Throw a coin and a die together 30 times and record your results in the table below.

Outcome	Tally	Frequency (f)	Relative Frequency $(\frac{f}{30})$
h1			
h2			
h3			
h4			
h5			
h6			
t1			
t2			
t3			
t4			
t5			
t6			
Sum			

1. Which of the 12 outcomes occurred most often?

   ————————————————————————————————

2. What is the sum of the first six fractions?

   ————————————————————————————————

3. What is the sum of the last six fractions?

   ————————————————————————————————

4. Is either of these two sums approximately equal to $\frac{1}{2}$?

   ————————————————————————————————

5. What is the sum of the two fractions

$$\frac{\text{Frequency for } h1}{30} + \frac{\text{Frequency for } t1}{30}?$$

_____

6. What is the sum of the two fractions

$$\frac{\text{Frequency for } h6}{30} + \frac{\text{Frequency for } t6}{30}?$$

_____

7. Is either of the two answers to questions 5 and 6 approximately equal to $\frac{1}{6}$?

_____

8. What is the difference of the two answers to questions 5 and 6?

_____

CONCLUSION: *Have you achieved the aim of the experiment? Explain.*

_____

_____

_____

_____

_____

_____

Elementary Probability 2

2.1 INTRODUCTION

In Chapter 1, the probability of an event A was defined as a fraction $\frac{n(A)}{n(S)}$. This fraction was obtained by first listing the outcomes, and then by counting the number of outcomes contained in the primitive space S and in the event A.

Examining this fraction more closely, we soon realize that all we really need to know in order to calculate the probability of an event A is the *number* of outcomes contained in A, that is, $n(A)$, and the *number* of outcomes contained in S, that is, $n(S)$. So, we would like to look for counting techniques and methods that would make the process of listing the outcomes unnecessary, if all we need to know is the probability of an event A.

In this chapter, we will examine the *multiplication principle*, a counting technique that enables us to find $n(S)$ and $n(A)$ without going through the process of listing all the outcomes. Furthermore, the graphical representation of this principle, called a *tree diagram*, gives us a systematic way of displaying the primitive sample space.

In addition to this principle, we will examine another method of calculating the probability of an event. That is, by using the known probability of another event, which is complementary to the event of interest. The multiplication principle and the complement method will extend considerably our ability to tackle more complicated, realistic, and interesting problems in probability.

Probability of an Event.

$$P(A) = \frac{n(A)}{n(S)}$$

Size of an event
We will call the number of outcomes in A the size of the event A. For example, the event $A = \{2,4,6\}$ has size 3, since $n(A) = 3$.

The multiplication principle is used: To find $n(S)$ and $n(A)$ without listing their outcomes.

Multiplication principle is also known as counting principle.

Complement Method will be studied in Sec. 2.4.

2.2 MULTIPLICATION PRINCIPLE

In Chapter 1, we found that the sample space for the coin-die experiment had a size of 12. We found this by listing the outcomes and counting them. This size of 12 could also be obtained without having to list all the outcomes. Remember that when we throw a

$n(S) = 12.$

A new look at the old experiment:
The act of throwing a coin and a die
is seen as consisting of two operations:
throwing a coin and throwing a die.

$A = \{h1,h2,h3,h4,h5,h6\}$

$n(A) = 6$

$n(S)$

coin and a die, the coin has two possible outcomes, $\{h,t\}$, and the die six possible outcomes, $\{1,2,3,4,5,6\}$. We know that for each of the 2 outcomes of the coin, we have 6 outcomes of the die to go with it. Therefore, given one particular outcome for the coin, we have 6 outcomes for the coin and the die together. Since the coin has 2 outcomes in all, we therefore must have

$$2 \times 6 = 12 \text{ possible outcomes}$$

for the coin-die experiment. This reasoning, which enables us to calculate the size of the primitive sample space without listing the outcomes, involves a very useful multiplication principle which underlies many probability problems that you are going to meet in this chapter.

Multiplication principle:
To use this principle, one must first
identify the two operations.

Multiplication Principle: *If an operation (or step) can result in* r *number of outcomes and after it has resulted in any one of these outcomes, a second operation (or step) can result in s number of outcomes, then these two operations* taken together *can result in r $\times$ s number of outcomes.*

Coin-die:

$r = 2$ outcomes

$s = 6$ outcomes

$r \times s = 12$ outcomes

We know that the coin-die experiment consists of two operations, the first being the flipping of a coin and the second the throwing of a die. (Note that we could also consider the throwing of a die as the first operation, and that of a coin as the second. Why?) In this case, we have $r = 2$, two outcomes for the flipping of a coin, and $s = 6$, six outcomes for the throwing of a die. Hence, the total number of outcomes for the two operations taken together, or for the coin-die experiment, is $2 \times 6 = 12$.

Tree diagram:
A display of a sample space.

Tree Diagram: *The multiplication principle for r = 2 and s = 6 as applied in the coin-die experiment can be illustrated by a diagram, called the tree diagram as shown in Figure 2.1.*

The tree diagram consists of line segments resembling the branches of a tree, and their intersecting point the joint of the branches. The tree diagram "grows" from left to right, and not from bottom to top like a real tree.

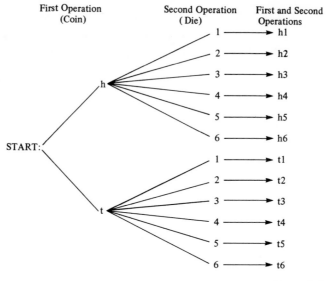

First Operation (Coin)   Second Operation (Die)   First and Second Operations

Figure 2.1

Each end-point of a line segment denotes an outcome of an operation, except the end-point marked with the word "Start." Branching off from the starting point, we have 2 end-points, one marked with *h*, heads, and the other with *t*, tails. Since the first operation (flipping the coin) has only 2 possible outcomes, we therefore have only 2 line segments coming from the starting point.

To represent the fact that there are 6 possible outcomes of the second operation (throwing the die) for *each* outcome of the first operation, we draw 6 line segments from *each* of the 2 outcomes, *h* and *t*, and mark their end-points by the numerals 1,2,3,4,5, and 6.

Now, to obtain the outcomes for the 2 operations *taken together*, we begin from the starting point, and move first along a line segment to *h*. This movement means that you have assumed that the first operation has resulted in a head. Then, move from *h* to 1 to obtain the outcome *h*1, from *h* to 2 to obtain the outcome *h*2, from *h* to 3 to obtain the outcome *h*3, from *h* to 4 to obtain the outcome *h*4, from *h* to 5 to obtain the outcome *h*5, and from *h* to 6 to obtain the outcome *h*6. As we read off each of these six outcomes, write it in the last column under "FIRST AND SECOND OPERATIONS."

How to read a tree diagram.

Determining the *outcomes* of the experiment.

6 outcomes for the coin and die when the coin comes up *h*.

What have we got so far? We have 6 outcomes of the *two operations taken together* where the outcome *h* of the first operation is fixed.

6 outcomes for the coin and die when the coin comes up *t*.

Returning now to the starting point, move along the line segment to *t*, and then to each of the six outcomes of the second operation, as before, to obtain another six outcomes for the two operations taken together. The column "FIRST AND SECOND OPERATIONS" now contains all the *12 possible outcomes* for the two operations taken together, that is, for the coin-die experiment.

Do we have *all* possible outcomes?

How do we know we have all the possible outcomes for the two operations taken together, that is, for the coin-die experiment? We know it because the procedure we adopted has covered all possible combinations of the line segments of the tree diagram in the order of the two operations such that *none is missing* and also *none is repeated*.

Note that the tree diagram in Figure 2.1, in addition to its graphical illustration for the multiplication principle, also provides us with *an organized way of listing the outcomes of the primitive sample space* for the experiment.

Examples:

1. *Two-die Experiment:* This experiment may be regarded as consisting of two operations: the *first operation* being the throwing of a green die, and the *second* the throwing of a red die. We could also consider the first operation being the throwing of a red die, and the second the throwing of a green die.

Each die has six possible outcomes, which is the same thing as saying that each operation has six possible outcomes. Since each outcome of the first operation can go with any one of the six outcomes of the second operation, we therefore can apply the multiplication principle for $r = 6$ and $s = 6$ to obtain the total number of outcomes for the two operations taken together. The total number of outcomes is $6 \times 6 = 36$, which can be interpreted as the size of the sample space for the two-die experiment.

Two dice:
$r = 6$
$s = 6$
$r \times s = 6 \times 6 = 36$

$n(S) = 36$ outcomes

The tree diagram in Figure 2.2 illustrates the multiplication principle for $r = 6$ and $s = 6$ involved in the two-die experiment.

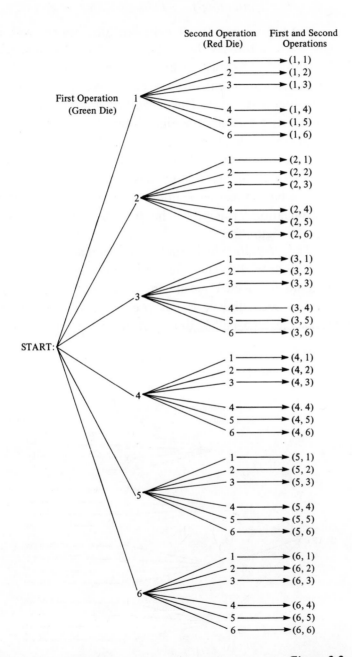

Figure 2.2

With replacement:
This means that the first ball is put back into the bag before the second ball is drawn.

Identify the two operations:
$r = 4$
$s = 4$
$r \times s = 4 \times 4 = 16$

$n(S) = 16$ outcomes

2. Experiment: *Drawing two balls in succession with replacement from a bag containing four balls of different colors.*

Let us assume the 4 balls are blue, green, yellow, and white, denoted respectively by *b, g, y,* and *w.* This experiment may be thought of as consisting of two operations: The *first operation* is the drawing of a ball from a bag of 4 balls, and thus it can result in 4 outcomes, $r = 4$. The *second operation* is the same as the first and thus $s = 4$, since the first drawn ball is *put back* into the bag before the second operation is performed.

We can apply the multiplication principle for $r = 4$ and $s = 4$ to obtain the total number of outcomes, which is $4 \times 4 = 16$ for the two operations taken together. (Supply the argument to justify the use of the multiplication principle). The tree diagram for this experiment is given in Figure 2.3.

The way in which the 16 outcomes have been listed implies that the tree diagram takes *order* into account. For example, the outcome *bg*, which means the green ball is drawn after the blue ball has been drawn, is *not the same* as the outcome *gb*, which means the blue ball is drawn after the green ball has been drawn.

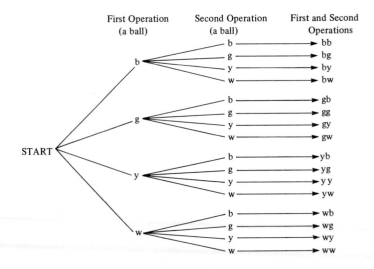

Figure 2.3

ORDER

This question of the order of operations is so important that we will pause here for a moment to consider it further.

In the experiment above we took two balls from the bag, but we took them in a special way. First, we took one ball, noted its color and put it back into the bag. Then we took a second ball, noted its color and put it back.

We have seen previously that the multiplication principle tells us that since there are four different colored balls, there are 16 possible outcomes. That is,

$$4 \times 4 = 16 \text{ outcomes.}$$

Using a tree diagram we saw that the sample space for this experiment is

$$S = \left\{ bb, bg, by, bw, gb, gg, gy, gw, yb, yg, yy, yw, wb, wg, wy, ww \right\}$$

The outcomes *bg* and *gb* are *different* because *bg* means "the blue ball was drawn first and then the green ball was drawn" while *gb* means just the opposite. The outcomes are different because the experiment asked us to take the *order* into account.

bg and *gb* are *different* outcomes.

But what if the experiment were worded somewhat differently?

Let us assume that we still had the same four balls in the bag. We are asked to draw two balls *together* from the bag. What would the sample space be in this case?

A different experiment

We could still apply the multiplication principle as before to get

$$4 \times 4 = 16 \text{ outcomes.}$$

We would also still draw the tree diagram as in Figure 2.3. However, in this case when we examine our outcomes we know, for example, that *bg* and *gb* are *not different* outcomes, but actually the *same outcome*. This is so because when we take the two balls *together* we know that we get a blue and a green ball—but we do not distinguish which one came first or second. That is, *we do not concern ourselves with the order of the selection.*

bg and *gb* are the *same* outcomes.

The order is not important.

bb cannot be an outcome.

Therefore, for this experiment, we exclude any outcome that is really the same as a previous one. (We also exclude those outcomes that cannot occur, such as *bb*.) So, we have

$$S = \left\{ \cancel{bb}, bg, by, bw, \cancel{gb}, \cancel{gg}, gy, gw, \cancel{yb}, \cancel{yg}, \cancel{yy}, yw, \cancel{wb}, \cancel{wg}, \cancel{wr}, \cancel{ww} \right\}$$

As a result, for the experiment of drawing two balls *together* from the bag, the sample space is

$$S = \left\{ bg, by, bw, gy, gw, rw \right\}$$

The Multiplication Principle takes into account the order in which operations occur.

If we use the principle in a situation where the order of the operations is not significant, we must exclude those outcomes that do not apply (as in the experiment above).

Without replacement:
This means that the first ball is *not* put back into the bag before the second ball is drawn.
Identify the two operations:

$y = 4$

$s = 3$

$y \times s = 4 \times 3 = 12$

$n(S) = 12$ outcomes

3. *Experiment: Drawing 2 balls (in succession) without replacement from a bag containing 4 balls of different colors.*

The two operations of this experiment are the same as those in the last example *except* that the first ball is *not* put back into the bag. Thus the second operation can result in only 3 outcomes, not 4 as in the last example. Hence, by applying the multiplication principle for $y = 4$ and $s = 3$, the total number of outcomes for this experiment is $4 \times 3 = 12$. These outcomes are displayed in Figure 2.4.

Note that the nature of the three outcomes of the second operation depends on which of the four outcomes was obtained in the first operation. For instance, if a blue ball, *b*, is first drawn, then the three possible outcomes left will be *g*, *y*, and *w*; and if a white ball, *w*, is first drawn, then the three outcomes left will be *b*, *g*, and *y*. However, the number of outcomes for the second operation is always three, irrespective of the actual outcome of the first operation.

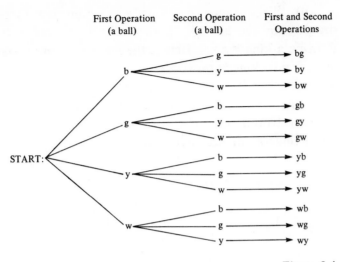

First Operation (a ball)	Second Operation (a ball)	First and Second Operations

Figure 2.4

4. *Experiment: From a deck of 52 cards, two cards are drawn in succession with replacement.*

a) How many different outcomes are possible?

b) How many outcomes are there consisting of an ace and a king?

c) What is the probability of drawing an ace and a king?

d) How many outcomes are there consisting of two aces?

e) What is the probability of drawing two aces?

Solution:

a) The first operation is drawing a card from a deck of 52 cards. Thus, the first operation can result in 52 outcomes, that is, $r = 52$. Since the first card drawn is replaced before the second draw, the second operation of drawing a card can also result in 52 outcomes, that is $s = 52$. Hence, by the multiplication principle, the two operations can result in

$$52 \times 52 = 2704 \text{ possible outcomes}$$

In other words, two cards can be drawn in 2704 ways. Note that the *order* in which the two cards are drawn is important.

Identify the two operations:
$r = 52$
$s = 52$

$r \times s = 52 \times 52$
$= 2704$

$n(S) = 2704$ outcomes

(3 of hearts, 5 of clubs) would be one outcome.

Identify the two operations:

$r = 4$

$s = 4$

$r \times s = 4 \times 4 = 16$ outcomes

(A of hearts, K of clubs) would be one outcome.

(K of clubs, A of hearts) would be one outcome.

$16 + 16 = 32$ outcomes

$$\frac{32}{2704} = \frac{2 \times 16}{169 \times 16}$$
$$= \frac{2}{169} \times \frac{16}{16}$$
$$= \frac{2}{169}$$

See Sec. A.4

(A of clubs, A of spades) would be one outcome.

b) Since there are four aces, then the first operation of drawing an ace is $r = 4$. In the same way, the next operation of drawing a king is $s = 4$. Hence, the number of outcomes consisting of an ace and a king *in that order* is

$$4 \times 4 = 16.$$

Now, if a king is drawn first, and then an ace, we have another 16 outcomes *in this order*. Hence, the number of outcomes consisting of an ace and a king *in any order* is

$$16 + 16 = 32.$$

c) From (b), the probability of drawing an ace and a king

$$= \frac{32}{2704}$$
$$= \frac{2}{169}$$

d) Since the first card drawn is put back into the deck before the second card is drawn, we have the same number of aces in either draw. So, the total number of possible outcomes consisting of two aces is $4 \times 4 = 16$.

e) From (d), the probability of drawing two aces

$$= \frac{16}{2704}$$
$$= \frac{1}{169}$$

PEANUTS

"IN DRIVING FROM TOWN A TO TOWN D YOU PASS FIRST THROUGH TOWN B AND THEN THROUGH TOWN C."

"IT IS 10 MILES FARTHER FROM A TO B THAN FROM B TO C AND 10 MILES FARTHER FROM B TO C THAN FROM C TO D. IF IT IS 390 MILES FROM A TO D, HOW FAR IS IT FROM A TO B?"

WELL, I KNEW IT WOULD HAPPEN SOONER OR LATER

WHAT'S THE MATTER?

MY EDUCATION HAS GROUND TO A HALT!

EXERCISES 2.2

1. A student in town A is going for a job interview in town C. He plans to go through town B on his trip to town C. The trip from town A to town B can be made in three different ways (bus, plane, train), and the trip from town B to town C in four different ways (bus, boat, plane, train). In how many ways can the entire trip be made from town A to town C?

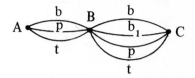

2. A committee consisting of a chairman and a secretary are to be chosen from a class of 20 students. Assuming that no one person can hold two offices on the same committee, how many committees can be chosen?

3. Bill has 5 shirts and 4 pairs of slacks. How many outfits consisting of a shirt and a pair of slacks can he select?

4. How many two-letter groups can be formed from the set of five letters $\{m,a,t,h,s\}$ if:

   a) No letters may be used more than once in a group?

   b) Repetition of letters in a group is allowed?

For example, mt is one group and tm is a different group.

5. There are five football teams in a town. How many matches must be arranged if every team must play every other team once.
(Check to be sure that your answer makes sense.)

6. A student is required to answer any 2 questions out of 5 questions in a math test. In how many ways can the student choose the two questions? (Check your answer.)

The multiplication principle takes into account *the order* in which the two operations are carried out. We can also use the principle to solve problems which do not require the consideration of order, as follows:

1. Solve these problems as if the order were important by the use of multiplication principle.

2. Exclude those outcomes that do not apply. (See explanation on Order, p. 69).

7. In the experiment of drawing two balls with replacement from a bag containing four balls (as discussed earlier in this section),

   a) What is the probability of getting one blue and one green ball?

   b) What is the probability of getting two blue balls?

   c) Are the two probabilities in (a) and (b) the same? Explain your answer.

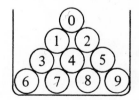

8. A bag contains 10 balls, each marked with a numeral from 0 through 9. Two balls are drawn in succession with replacement.

   a) How many outcomes are possible?

   b) What is the probability of drawing two balls with the same numeral?

What is $n(S)$?

Picture cards:

Suit
{club, spade, diamond, heart}

9. In problem 8, how many outcomes are possible if the two balls are drawn in succession *without* replacement?

   a) What is the probability of drawing two balls with the same numeral?

10. a) How many different two-digit numbers can be formed from the set of digits {0,1,2,3,4,5,6,7,8,9},

   i) if repetition of digits is allowed?

   ii) if repetition of digits is not allowed?

   b) What is the probability that a two-digit number consisting of two different digits will be randomly chosen from the set of all two-digit numbers, {00, 01, . . . , 99}?

11. From a deck of 52 cards, *two* cards are drawn in succession without replacement.

   a) How many different outcomes are possible?

   b) How many of these outcomes consist of picture cards only?

   c) What is the probability of drawing two picture cards only?

   d) What is the probability of drawing two cards of hearts?

   e) What is the probability of drawing two cards of the same suit?

2.3 EXTENDED MULTIPLICATION PRINCIPLE

We have defined the multiplication principle for two operations. It also applies to experiments which may be regarded as consisting of three or more operations. Some simple examples are those experiments of flipping 3 coins, 4 coins, 5 coins, and so on. Here, we shall confine our discussion to experiments consisting of three operations only.

Examples

1. **Three-coin Experiment:** *Flipping a penny, a nickel, and a dime.*

The three operations that make up this experiment are: the flipping of a penny, the flipping of a nickel, and the flipping of a dime. In order to apply the multiplication principle as stated in Section 2.2, the three operations must be reduced to two. This is done by treating the first two operations as one single operation,

Experiments having three or more operations

1. Identify the three operations.

2. Reduce three operations to two.

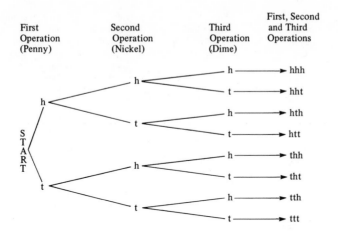

First Operation (Penny)	Second Operation (Nickel)	Third Operation (Dime)	First, Second and Third Operations

Figure 2.5

which can result in 4 outcomes (why 4?), and the third operation, leaving it as it is, as the second operation.

Then, the first operation (flipping a penny *and* a nickel) can result in 4 outcomes ($r = 4$). Each of these 4 outcomes can be associated with either of the 2 outcomes of the second operation (flipping the dime, $s = 2$). Hence, the total number of outcomes for the two operations taken together is $4 \times 2 = 8$. This is the total number of outcomes for the three operations, or for the three-coin experiment. The tree diagram is shown in Figure 2.5. Note that *each outcome* in this experiment consists of *three parts*. For example, the outcome *hth* has *h* (a head on the penny), *t* (a tail on the nickel), and *h* (a head on the dime).

Try to draw the tree diagram for the experiment of flipping 4 coins.

2. Experiment: *Drawing 3 balls (in succession) without replacement from a bag containing 4 balls of different colors.*

As in the last example, we regard the first two operations of drawing two balls with no replacement as one single operation with $4 \times 3 = 12$ possible outcomes. Hence, by using the multiplication principle for $r = 12$ and $s = 2$, we have the total number of outcomes for the three operations, which is $12 \times 2 = 24$. The tree diagram for this experiment is given in Figure 2.6.

3 coins:
$$r = 4$$
$$s = 2$$
$$r \times s = 4 \times 2$$
$$= 8 \text{ outcomes}$$

hth is one outcome.

Without replacement

1. Identify the three operations.
2. Reduce three operations to two.

$$r = 12$$
$$s = 2$$
$$r \times s = 12 \times 2$$
$$= 24 \text{ outcomes}$$

First Operation (a ball)	Second Operation (a ball)	Third Operation (a ball)	First, Second and Third Operations
		y	bgy
	g	w	bgw
		g	byg
b	y	w	byw
		g	bwg
	w	y	bwy
		y	gby
	b	w	gbw
		b	ygb
g	y	w	ygw
		b	gwb
	w	y	gwy
		g	ybg
	b	w	ybw
		b	ygb
y	g	w	ygw
		b	ywb
	w	g	ywg
		g	wbg
	b	y	wby
		b	wgb
w	g	y	wgy
		b	wyb
	y	g	wyg

Figure 2.6

3.   **Experiment**: *Chuck-a-luck (Big Six)*.

Three dice (one green, one red, and one white) are rolled once.
a)   How many different outcomes are possible?
b)   How many outcomes are there consisting of three 6's?
c)   What is the probability of rolling three 6's?
d)   How many outcomes are there consisting of two 6's?
e)   What is the probability of rolling two 6's?
f)   How many outcomes are there consisting of one 6?
g)   What is the probability of rolling one 6?

Solution:

a)   Since we roll three dice, we have outcomes consisting of three parts, such as (3,5,2). The numeral 3 denotes the outcome of the green die, the numeral 5 the outcome of the red die, and the numeral 2 the outcome of the white die.

We already know that the green and the red dice can result in 36 outcomes, and that the white die can result in 6 outcomes. Thus, the rolling of three dice, by the multiplication principle, can result in 36 × 6 = 216 outcomes. That is $n(S) = 216$.
(See Figure 2.7 for the 216 outcomes.)

b)   In order to get three 6's, each and every one of the three dice must come up 6. So, there is only one outcome, namely (6,6,6), that can result in three 6's.

c)   From (b), the probability of rolling three 6's

$$= \frac{1}{216}, \text{ since } n(S) = 216$$

d)   We note that two 6's can occur in three separate ways: (6,6,?), (6,?,6), (?,6,6). Let us consider the first way: (6,6,?). Since we want only two 6's, the question mark can be replaced by any one of the five numerals 1,2,3,4, and 5. Thus, this way can result in 5 outcomes with two 6's: (6,6,1), (6,6,2), (6,6,3), (6,6,4) and (6,6,5).

Similarly, the second way can result in 5 possible outcomes with only two 6's, and the same for the third way. Therefore, the total number of outcomes with only two 6's is 5 + 5 + 5 = 15. (See Figure 2.7 for the 15 outcomes).

e)   From (d), the probability of rolling only

$$\text{two 6's} = \frac{15}{216}$$
$$= \frac{5}{72}$$

f)   Here, we again have three different ways of getting only one 6. (6,?,?), (?,6,?), (?,?,6). Let us consider the first way: (6,?,?). Since

Method:
1. Identify the three operations.
2. Reduce three operations to two.
   $r = 36$
   $s = 6$
   $r \times s = 36 \times 6$
   $= 216$ outcomes

(6,1,6), (6,2,6), etc.
(1,6,6,), (2,6,6,), etc.
15 outcomes like these.

$$\frac{15}{216} = \frac{5 \times 3}{72 \times 3}$$
$$= \frac{5}{72} \times 1$$
$$= \frac{5}{72}$$

Sample Space For 3-die Experiment

(1,1,1)	(1,1,2)	(1,1,3)	(1,1,4)	(1,1,5)	(1,1,6)
(1,2,1)	(1,2,2)	(1,2,3)	(1,2,4)	(1,2,5)	(1,2,6)
(1,3,1)	(1,3,2)	(1,3,3)	(1,3,4)	(1,3,5)	(1,3,6)
(1,4,1)	(1,4,2)	(1,4,3)	(1,4,4)	(1,4,5)	(1,4,6)
(1,5,1)	(1,5,2)	(1,5,3)	(1,5,4)	(1,5,5)	(1,5,6)
(1,6,1)	(1,6,2)	(1,6,3)	(1,6,4)	(1,6,5)	(1,6,6)
(2,1,1)	(2,1,2)	(2,1,3)	(2,1,4)	(2,1,5)	(2,1,6)
(2,2,1)	(2,2,2)	(2,2,3)	(2,2,4)	(2,2,5)	(2,2,6)
(2,3,1)	(2,3,2)	(2,3,3)	(2,3,4)	(2,3,5)	(2,3,6)
(2,4,1)	(2,4,2)	(2,4,3)	(2,4,4)	(2,4,5)	(2,4,6)
(2,5,1)	(2,5,2)	(2,5,3)	(2,5,4)	(2,5,5)	(2,5,6)
(2,6,1)	(2,6,2)	(2,6,3)	(2,6,4)	(2,6,5)	(2,6,6)
(3,1,1)	(3,1,2)	(3,1,3)	(3,1,4)	(3,1,5)	(3,1,6)
(3,2,1)	(3,2,2)	(3,2,3)	(3,2,4)	(3,2,5)	(3,2,6)
(3,3,1)	(3,3,2)	(3,3,3)	(3,3,4)	(3,3,5)	(3,3,6)
(3,4,1)	(3,4,2)	(3,4,3)	(3,4,4)	(3,4,5)	(3,4,6)
(3,5,1)	(3,5,2)	(3,5,3)	(3,5,4)	(3,5,5)	(3,5,6)
(3,6,1)	(3,6,2)	(3,6,3)	(3,6,4)	(3,6,5)	(3,6,6)
(4,1,1)	(4,1,2)	(4,1,3)	(4,1,4)	(4,1,5)	(4,1,6)
(4,2,1)	(4,2,2)	(4,2,3)	(4,2,4)	(4,2,5)	(4,2,6)
(4,3,1)	(4,3,2)	(4,3,3)	(4,3,4)	(4,3,5)	(4,3,6)
(4,4,1)	(4,4,2)	(4,4,3)	(4,4,4)	(4,4,5)	(4,4,6)
(4,5,1)	(4,5,2)	(4,5,3)	(4,5,4)	(4,5,5)	(4,5,6)
(4,6,1)	(4,6,2)	(4,6,3)	(4,6,4)	(4,6,5)	(4,6,6)
(5,1,1)	(5,1,2)	(5,1,3)	(5,1,4)	(5,1,5)	(5,1,6)
(5,2,1)	(5,2,2)	(5,2,3)	(5,2,4)	(5,2,5)	(5,2,6)
(5,3,1)	(5,3,2)	(5,3,3)	(5,3,4)	(5,3,5)	(5,3,6)
(5,4,1)	(5,4,2)	(5,4,3)	(5,4,4)	(5,4,5)	(5,4,6)
(5,5,1)	(5,5,2)	(5,5,3)	(5,5,4)	(5,5,5)	(5,5,6)
(5,6,1)	(5,6,2)	(5,6,3)	(5,6,4)	(5,6,5)	(5,6,6)
(6,1,1)	(6,1,2)	(6,1,3)	(6,1,4)	(6,1,5)	(6,1,6)
(6,2,1)	(6,2,2)	(6,2,3)	(6,2,4)	(6,2,5)	(6,2,6)
(6,3,1)	(6,3,2)	(6,3,3)	(6,3,4)	(6,3,5)	(6,3,6)
(6,4,1)	(6,4,2)	(6,4,3)	(6,4,4)	(6,4,5)	(6,4,6)
(6,5,1)	(6,5,2)	(6,5,3)	(6,5,4)	(6,5,5)	(6,5,6)
(6,6,1)	(6,6,2)	(6,6,3)	(6,6,4)	(6,6,5)	(6,6,6)

Figure 2.7

(6,1,1), (6,1,2), etc.
25 outcomes like these

(1,6,1), (1,6,2), etc.
25 outcomes like these

we want only one 6, neither of the question marks can be replaced by the numeral 6. So, we can use only one of the five numerals 1,2,3,4, and 5, to replace the first question mark, and also one of the five to replace the second question mark. So, we have $5 \times 5 = 25$ ways of replacing the two question marks by the five numerals. In other words, the first way can result in 25 outcomes.

Similarly, the second and the third ways can each result in 25 outcomes.

Therefore, the total number of outcomes with only one 6 is

$$25 + 25 + 25 = 75.$$

(See Figure 2.7 for the 75 outcomes.)

g) From (f), the probability of rolling only

$$\text{one } 6 = \frac{75}{216}$$
$$= \frac{25}{72}$$

(We will complete the discussion of this experiment in Chapter 3.)

Note that this example also shows that the multiplication principle is easier to use than the tree diagram, especially when the size of a primitive sample space or an event is large.

EXERCISES 2.3

1. If a game can result in any one of three outcomes: win, loss, or tie, then how many different outcomes are possible in a series of 3 games?

2. A girl has 9 dresses, 5 hats, and 6 pairs of shoes. How many different outfits can she select where each outfit consists of a dress, a hat, and a pair of shoes?

3. In Bill's personal library, there are 10 mathematics books, 15 science books, and 20 novels. In how many ways can Bill choose 3 books, one of each kind?

4. In how many ways can a committee of three consisting of a chairman, a secretary and a treasurer be chosen from a class of 30 students assuming that no one person can hold more than one of these offices?

5. A short test consists of four multiple-choice questions, each of which has five options. In how many ways can a person guess the answers to the four multiple-choice questions?

Method:
1. Identify all the operations.
2. Multiply the number of times each of the operations can be performed.

There are four operations.

How many letters in the alphabet? How many digits in our decimal system?

What does each of these conditions mean?

$$P(A) = \frac{n(A)}{n(S)}$$

Experiment VI, p.82

Pascal's Triangle

Can you use these numbers to find the answers to problems 11 and 12?

6. A person throws four dice: one green, one red, one white, and one yellow. How many different outcomes are possible?

7. How many different license plates can be made if each plate is to consist of a letter of the alphabet followed by three digits?

8. In exercise 7, how many different license plates are possible if the first digit cannot be zero, and a digit cannot be repeated?

9. How many *three*-digit numbers can be formed from the set of ten decimal digits:
   a) Without repetition?
   b) With repetition?
   c) With repetition, but where the digit zero cannot be used as the first digit?
   d) What is the probability that a randomly chosen three-digit number consists of three different digits?

10. How many *four*-digit numbers can be formed from the set of ten digits if:
   a) Repetition is allowed?
   b) If repetition is not allowed?
   c) What is the probability that a randomly chosen four-digit number consists of four different digits?

(Note that the answer given to (c) provides the key to the prediction, close to 50, in the telephone experiment at the end of this Section 2.3.)

11. In the three-coin experiment, what is the probability of getting:
   a) 3 heads?
   b) 2 heads?
   c) 1 head?
   d) no heads?

12. In a four-coin experiment (penny, nickel, dime, and quarter):
   a) How many possible outcomes are there?
   b) What is the probability of getting 2 heads and 2 tails?
   c) What is the probability of getting 3 heads and 1 tail?

13. The policeman and the thief:

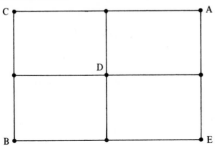

In this figure, assume that the horizontal lines are streets and the vertical lines are avenues. There is a policeman at the street corner A and a thief at the street corner B, and they are equally likely to take either way at a street corner.

   If they simultaneously walk two blocks from their beginning positions *A* and *B*, then each must stop at one of the street corners *C, D,* or *E.*

a)   What is the probability that they will stop at the same street corner?

(Hint: First figure out all the possible final positions of the policeman and the thief. That is, the primitive sample space for this "experiment.")

b)   Is this problem mathematically the same as flipping 4 coins as given in Exercise 12?

Name: _____

Experiment VI: *Last four digits of 100 telephone numbers.*

AIM: *To determine whether each of the four digits is chosen from the ten decimal digits 0 to 9 with equal probability.*

A. Take any telephone directory and open it to any page. Count off the first 100 telephone numbers. Next, count the number of telephone numbers whose last four digits are all different. (The number 2701 is an example but not the number 2214).

1. How many of the 100 numbers have all four digits different?

    _____

2. Is your answer close to 50?

    _____

3. If yes, explain why.

    _____

    _____

4. If no, suggest reasons why your answer is not close to 50. (You may want to re-examine the set of 100 telephone numbers that you chose for the explanation.)

    _____

    _____

    _____

    _____

B. Now, collect all numbers so obtained by students in your class. Add up these numbers and then divide their sum by the number of students who provided these numbers.

5. What is your new answer?

   _____

6. Is your new answer close to 50?

   _____

7. Is your new answer closer to 50 than your old answer given in question 1? Explain.

   _____

   _____

   _____

   _____

CONCLUSION: *Have you achieved the aim of the experiment? Explain.*

   _____

   _____

   _____

   _____

   _____

   _____

$\overline{A}$

"not A"

Complement of *A*

STOP

A and $\overline{A}$

no common outcomes

Describe event *A* in words.

2.4 COMPLEMENT OF AN EVENT

In carrying out an experiment, the object is usually to find out whether some event A will occur. In doing so, we "ignore" an event which contains all the outcomes in S except those that are contained in A. For example, in the die experiment, we investigated $A = \{2,4,6\}$ and "ignored" the event $\{1,3,5\}$, which contains all the outcomes in $S = \{1,2,3,4,5,6\}$ except those in A. We call the event $\{1,3,5\}$ the *complement* or *complementary event* of the event *A*, and indicate its relationship to *A* by using the symbol $\overline{A}$ to denote it. The symbol "$\overline{A}$" is read "*A* bar" or "not *A*."

Definition. *If* A *is an event of a sample space* S, *then the* complement *or* complementary event *of* A *is the event which contains all the outcomes in the sample space* S *that are* not *contained in* A.

From this definition, we can see that the number of outcomes in $\overline{A}$ is equal to the number of outcomes in *S minus* the number of outcomes in *A*. That is,

$$n(\overline{A}) = n(S) - n(A).$$

Moreover, the two events A and $\overline{A}$ do not contain any outcomes that are the same. In other words, *A and $\overline{A}$ have no common outcomes.*

Examples:

1. **Three-coin experiment:** *Flipping a penny, a nickel and a dime.*

Recall that $S = \{hhh,\ hht,\ hth,\ thh,\ tth,\ tht,\ htt,\ ttt\}$ and $n(S) = 8$.

    a) Let $A = \{hht,\ hth,\ thh\}$.

      Then $\overline{A} = \{hhh,\ tth,\ tht,\ htt,\ ttt\}$.

Check:

$$n(\overline{A}) = n(S) - n(A) = 8 - 3 = 5,$$

and A and $\overline{A}$ have no common outcomes.

b) Let $B = \{hhh,\ ttt\}$.

Describe event B in words.

Then $\overline{B} = \{hht,\ hth,\ thh,\ tth,\ tht,\ htt\}$.

Check:

$$n(\overline{B}) = n(S) - n(B) = 8 - 2 = 6,$$

and also B and $\overline{B}$ have no common outcomes.

2. Experiment: *Drawing two balls in succession* with *replacement from a bag containing four balls of different colors (blue, green, red, and white).*

With replacement

Recall that

$$S = \{bb,\ bg,\ br,\ bw,\ gb,\ gg,\ gr,\ gw,\ rb,\ rg,\ rr,\ rw,\ wb,\ wr,\ wg,\ ww\},$$
$$\text{and } n(S) = 16$$

a) Let $C = \{bb,\ gg,\ rr,\ ww\}$.

Describe event C in words.

Then $n(C) = 4$, and

$\overline{C} = \{bg,\ br,\ bw,\ gb,\ gr,\ gw,\ rb,\ rg,\ rw,\ wb,\ wr,\ wg\}$.

Check that the outcomes in $\overline{C}$ are correctly listed.

b) Let $D = \{bg,\ gb,\ gg,\ gr,\ gw,\ rg,\ wg\}$

Describe event D in words.

Then $n(D) = 7$ and

$\overline{D} = \{bb,\ br,\ bw,\ rb,\ rr,\ rw,\ wb,\ wr,\ ww\}$.

Check that the outcomes in $\overline{D}$ are correctly listed.

Before we consider another example, let us take a look at the verbal descriptions of C and $\overline{C}$, and D and $\overline{D}$ in the last example.

Event	Verbal Description
C	Two balls of the *same* color.
$\overline{C}$	Two balls of *different* colors.
D	*At least* one of the two balls is green.
$\overline{D}$	*Neither* of the two balls is green. (Or two non-green balls).

Opposite Meanings.

Thus, we may say that the verbal descriptions of an event and its complement have *opposite meanings*.

3. *In the last experiment, find:*

    a)   The probability of getting two balls of different colors, or $P(\overline{C})$.

    b)   The probability of getting two non-green balls, or $P(\overline{D})$.

Solution:

There are two methods of calculating $P(\overline{C})$ and $P(\overline{D})$: The direct and the indirect methods.

Direct Method:

Direct method of finding $P(\overline{C})$.

$$P(\overline{C}) = \frac{n(\overline{C})}{n(S)}$$

    a)   Since $\overline{C} = \{bg,\ br,\ bw,\ gb,\ gr,\ gw,\ rb,\ rg,\ rw,\ wb,\ wr,\ wg\}$ and $n(\overline{C}) = 12$,

        then $P(\overline{C}) = \frac{12}{16} = \frac{3}{4}$

    b)   Since $\overline{D} = \{bb,\ br,\ bw,\ rb,\ rr,\ rw,\ wb,\ wr,\ ww\}$ and $n(\overline{D}) = 9$,

        then $P(\overline{D}) = \frac{9}{16}$

Find $\overline{C}$ first.

In this method, we need to find $\overline{C}$ and $\overline{D}$ first before we can calculate their probabilities.

Indirect Method:

a) Since $n(C) = 4$, then

$$P(\overline{C}) = \frac{n(\overline{C})}{n(S)}$$

$$= \frac{n(S)-n(C)}{n(S)} \text{, since } n(\overline{C}) = n(S) - n(C)$$

$$= \frac{16-4}{16}$$

$$= \frac{12}{16}$$

$$= \frac{3}{4}$$

b) Since $n(D) = 7$, then

$$P(\overline{D}) = \frac{n(\overline{D})}{n(S)}$$

$$= \frac{n(S)-n(D)}{n(S)}$$

$$= \frac{16-7}{16}$$

$$= \frac{9}{16}$$

In this method, we do not need to find $\overline{C}$ and $\overline{D}$ first. We use the formula $n(\overline{A}) = n(S) - n(A)$ in the calculation of their probabilities.

Question:

Which of the two methods should be used for calculating $P(\overline{A})$?

It depends on which of the two methods can be more readily used for calculating $P(\overline{A})$.

The formula $P(\overline{A}) = \frac{n(S)-n(A)}{n(S)}$ can also be expressed as:

$$P(\overline{A}) = \frac{n(S)}{n(S)} - \frac{n(A)}{n(S)}, \text{ Why?}$$

That is, $P(\overline{A}) = 1 - P(A)$

Indirect method of finding $P(\overline{C})$.

It is not necessary to find $\overline{C}$ first.

Representations of a number:
A number can be represented in many forms.
Examples:
$1 = \frac{16}{16} = \frac{36}{36} = 100\%$, etc.

$\frac{1}{2} = \frac{2}{4} = \frac{3}{6} = 50\%$, etc.

$6 = 3 + 3 = 7 - 1 = \frac{6}{1}$, etc.

The choice of the representation is a matter of convenience.

See Sec. A.6, A.7

$\frac{n(S)}{n(S)} = 1$

$$P(\overline{A}) = 1 - P(A)$$

This formula expresses the relationship of an event and its complement in terms of probabilities. It is a very useful formula to remember.

Examples:

1. In the two-die experiment, find:

   a) the probability of *not* getting a double.

   b) the probability of getting a sum less than 11.

See Sample Space, p. 10

Solution:

a) Since the probability of getting a double is $\frac{6}{36}$, then the probability of not getting a double is

$$1 - \frac{6}{36} = \frac{30}{36} = \frac{5}{6}.$$

$$1 - \frac{6}{36} = \frac{36}{36} - \frac{6}{36}$$
$$= \frac{36 - 6}{36}$$
$$= \frac{30}{36}$$
$$= \frac{5 \times 6}{6 \times 6}$$
$$= \frac{5}{6} \times 1$$
$$= \frac{5}{6}$$

Why?

b) Let A be the event that the sum of the two dice is *equal to* or *greater than* 11. That is, $A = \{(5,6), (6,5), (6,6)\}$.

Thus, the desired probability

$$= 1 - P(A)$$
$$= 1 - \frac{3}{36}$$
$$= \frac{11}{12}$$

See Sec. A.5

Note that we have used the formula $P(\overline{A}) = 1 - P(A)$ for (a) and (b), since it is more convenient to use to find the desired probabilities.

2. *Birthday Problem:* Assume that there are 365 days in a year and that a person can have any one of these 365 days as his birthday.

a) If two persons are randomly chosen, what is the probability that they both have the same birthday (same day of the year)?

Solution:

a) Since a person can have any one of the 365 days as his birthday, then his birthday can result in 365 ways. By the *multiplication principle* the two persons' birthdays can result in 365 × 365 ways. That is, $n(S) = 365 \times 365$.

Now, if the second person's birthday is to be different from the first person's, then his birthday can result in only 364 ways. Hence, the two persons can have *different birthdays* in 365 × 364 ways.

That is, $n(A) = 365 \times 364$, where A is the event that the two persons have different birthdays. Thus $\overline{A}$ is the event that the two persons have the *same* birthday.

Using the formula, $P(\overline{A}) = 1 - P(A)$, we have

$$P(\overline{A}) = 1 - \frac{365 \times 364}{365 \times 365}$$

$$= \frac{1}{365}$$

$$\approx .003 \text{ (very small indeed).}$$

$$1 = \frac{365}{365}$$

$$365\overline{)\begin{array}{r} .0027 \\ 1.0000 \\ \underline{730} \\ 2700 \\ \underline{2555} \\ 145 \end{array}}$$

$$\frac{1}{365} \approx .003$$

See Sec. B3, B4

b) If three persons are randomly chosen, what is the probability of *at least* two of them having the same birthday?

Solution:

Repeating the reasoning given in (a), we know that the three persons' birthdays can result in 365 × 365 × 365 ways. That is, $n(S) = 365 \times 365 \times 365$. Furthermore, the three persons can have different birthdays in 365 × 364 × 363 ways. That is, $n(A) = 365 \times 364 \times 363$.

By permission of John Hart and Field Enterprises Inc.

$$1 = \frac{365 \times 365}{365 \times 365}$$

$$
\begin{array}{r}
.0082 \\
133225 \overline{)1093.0000} \\
1065\ 800 \\
\hline
27\ 2000 \\
26\ 6450 \\
\hline
5550
\end{array}
$$

See Sec. B.3

Therefore,

$$P(\overline{A}) = 1 - \frac{365 \times 364 \times 363}{365 \times 365 \times 365}$$

$$= 1 - \frac{364 \times 363}{365 \times 365}$$

$$= \frac{1093}{133,225}$$

$$\approx .008 \text{ (very small)}$$

Question:

How many persons do you think we will need to have before we can have the *probability of 0.5* or more that at least two persons have the same birthday?

Take a guess!

Write down your guess, _____

Question:

How many persons do you think we will need to have before we can be *almost certain* that at least two persons have the same birthday?

Take a guess!

Write down your guess, _____

Now check your guesses with the table on the facing page, which is calculated by the same method we have just discussed in the last example.

Are you surprised to note that we need only 23 persons to have a 50% probability that at least two persons in the group have the same birthday?

Are you also surprised to note we need only 41 persons to have a 90% probability that at least two persons in the group have the same birthday?

Why not try this birthday experiment in your class?

Number of Persons	Approximate probability that at least two persons have the same birthday
2	.003
3	.008
4	.016
8	.074
12	.167
16	.284
20	.411
22	.476
23	.507
24	.538
28	.654
32	.753
40	.891
48	.960
56	.988
64	.997

EXERCISES 2.4

1. What is the complement of S?

2. What is the complement of ϕ?

3. In the three-coin experiment, if $C = \{hhh,\ hth,\ ttt\}$,

    a) Find $\overline{C}$.

    b) Find the complement of $\overline{C}$.

4. What is the complement of $\overline{A}$?

5. If $n(S) = 30$ and $n(A) = 12$, what is $n(\overline{A})$?

6. Is $n(S) = n(A) + n(\overline{A})$? Explain.

7. If $P(A) = \frac{4}{13}$, what is $P(\overline{A})$?

8. Is $P(A) + P(\overline{A}) = 1$? Explain.

9. In the two-coin experiment, find:

    a) The probability of getting *at least* one head.

    b) The probability of getting *no heads*.

See sample space, p. 74

$$\overline{\overline{A}}$$

The complement of the complement of A

$$\overline{\overline{A}} = A$$

See sample space p. 53

See sample space p. 53

$$P(A) = \frac{n(A)}{n(S)}$$

10. In the two-die experiment, find:

   a) The probability of getting a sum of 7 on the two dice.

   b) The probability of *not* getting a sum of 7 on the two dice.

11. In a win-loss game, if Bob's probability of winning the game is .4, what is Bob's probability of losing the game?

12. A bag contains a set of 100 slips of paper which bear the numbers from 0 to 99. If three slips are randomly drawn in succession *with* replacement.

   a) What is the probability that the three numbers drawn are all different?

   b) What is the probability that at least two of the three numbers are the same?

13. Using the birthday table given on page 91, find:

   a) The probability that none of the people in a group of 20 have the same birthday.

   b) The probability that none of the people in a group of 48 have the same birthday.

14. Is it likely that no two of the 38 Presidents of the United States in the period 1789-1975 had the same birthday? Explain your answer.

SUMMARY

Experiments:

1. *Drawing two cards in succession with replacement from a deck of 52 cards*

$$n(S) = 52 \times 52 = 2704 \text{ outcomes}$$

One of these outcomes is $(5_h, k_c)$, which means 5 of hearts, and king of clubs.

2. *Drawing two cards in succession without replacement from a deck of 52 cards*

$$n(S) = 52 \times 51 = 2652 \text{ outcomes}$$

One of these outcomes is $(2_d, j_s)$, which means 2 of diamonds, and jack of spades.

3. *Three-coin Experiment: Flipping a penny, a nickel and a dime together*

$$n(S) = 2 \times 2 \times 2 = 8 \text{ outcomes}$$

One of these outcomes is (h, h, t), which means heads on penny, heads on nickel, and tails on dime.

4. *Three-die Experiment: Rolling one green, one red, and one white dice together*

$$n(S) = 6 \times 6 \times 6 = 216 \text{ outcomes}$$

One of these outcomes is $(3, 5, 2)$, which means 3 on green, 5 on red, and 2 on white die.

Terms and Symbols:

Terms	*Symbols*	*Page*
Size of an event A	$n(A)$	63
Multiplication principle	$r \times s$	63
Extended multiplication principle	$r \times s \times t \times \ldots$	74
Tree diagram		64
Complement of A (Complementary event of A)	$\overline{A}$	84

Formula:

1. $n(\overline{A}) = n(S) - n(A)$, where $A < S$
   or $n(S) = n(A) + n(\overline{A})$
2. $P(\overline{A}) = 1 - P(A)$
   or $P(\overline{A}) + P(A) = 1$

REVIEW EXERCISES.

1. Find the size of the *primitive* sample space for each of the following chance experiments.

   a)   Four dice are rolled together once.

   b)   Five dice are rolled together once.

   c)   Two coins and three dice are thrown together once.

   d)   Three cards are drawn *without* replacement from a deck of 52 cards.

   e)   Four cards are drawn *without* replacement from a deck of 52 cards.

2. State three properties of complementary events A and $\bar{A}$, and give an example to illustrate each property.

3. Let the primitive sample space for an experiment be
   $$S = \{0, 1, 2, 3, 4, 5, 6, 7, 8, 9\}.$$

   a)   Give an event A such that $n(A) = 2$.

   b)   Give an event B such that $A \subset B$ and $n(B) > n(A)$.

   c)   What are the outcomes in $\bar{A}$?

   d)   What are the outcomes in $\bar{B}$?

   e)   Is $\bar{A} \subset \bar{B}$?

   f)   Is $\bar{B} \subset \bar{A}$?

4. Let $n(S) = 90$, $n(A) = 35$, $n(B) = 78$, and $n(C) = 17$.

   a)   Find $n(\bar{A})$, $n(\bar{B})$, and $n(\bar{C})$

   b)   Find $P(\bar{A})$, $P(\bar{B})$, and $P(\bar{C})$

   c)   Is $P(\bar{C}) > P(C)$?

   d)   Is $P(\bar{B}) > P(B)$?

5. Let $n(S) = 125$, $n(\bar{D}) = 47$, $n(\bar{E}) = 112$, and $n(\bar{F}) = 97$.

   a)   Find $n(D)$, $n(E)$, and $n(F)$

   b)   Find $P(D)$, $P(E)$, and $P(F)$

   c)   Which of the three probabilities given in (b) is the largest?

6. In the three-coin experiment,

See sample space p. 74

   a)   Give two events X and Y such that $P(X) + P(Y) = 1$.

   b)   Are X and Y complementary events?

   c)   Which of the two events is more likely to occur? Why?

7. a)   How many 5-digit numbers can be formed from the set of five digits $\{1, 2, 3, 4, 5,\}$ if no digits can be used more than once?

   b)   How many of these numbers are even?

   c)   What is the probability of getting an even number?

   d)   What is the probability of getting an odd number?

8. A license plate is made by using a letter of the alphabet followed by three digits (for example, H206).

   a) How many different license plates can be made? (The ten decimal digits can be used and repetition is allowed.)

   b) How many of these plates have the form of a letter followed by the same digit three times (for example, C666)?

   c) What is the probability of getting a license plate having the arrangement in (b)?

9. Three dice are rolled together once.

   a) Let $H = \{$no 6's on one roll of the three dice$\}$. Give a verbal description for the complement of the event H.

   b) What is the probability of getting no 6's?
   (Hint: Use the results of (a) and Example 3 of Section 3.3)

10. Five-card stud poker game.

   This table gives the number and the probability to six decimal places of each of 10 poker hands, with royal flush being ranked the highest and no pair the lowest.

Poker Hand		Number	Probability (to six decimal places)
(I)	Royal flush	4	.000,002
(II)	Straight flush	36	.000,014
(III)	Four of a kind	624	.000,240
(IV)	Full house	3,744	.001,441
(V)	Flush	5,108	.001,965
(VI)	Straight	10,200	.003,925
(VII)	Three of a kind	54,912	.021,130
(VIII)	Two pairs	123,552	.047,539
(IX)	One pair	1,098,240	.422,569
(X)	No pair	1,302,540	.501,177
	TOTAL	2,598,960	1.000,002

   a) Write in words the number of each of the poker hands given in (VI) – (X).

   See Sec. B.1.

   b) Write in words the probability of each of the poker hands given in (1) – (V).

c) Why is the total of all the probabilities slightly greater than one?

d) Which of the poker hands is most likely to occur? Give its probability in percent.

e) Which of the poker hands is least likely to occur? Give its probability in percent.

f) What is the approximate probability in percent that a poker hand is not higher in rank than one pair?

g) What is the approximate probability in percent that a poker hand is higher in rank than one pair?

Mathematical Expectations 3

PROBLEM OF POINTS*

One day, Bob and Joe decided to flip a penny a number of times. Bob chose heads, and Joe tails. They each put up $10, and agreed that the $20 would be taken by the person who first won 5 games.

They took turns flipping the penny. After 7 flips, Bob had won 4 games and Joe 3 games. It was now Joe's turn to flip the penny. But, he decided to call off the bet.

"Okay, the twenty dollars is mine," said Bob.

"No, you haven't won the bet," Joe answered, "We should each take back our own $10."

"That's not fair!"

"Why not?" Joe asked.

"I've won more games than you," answered Bob with some dissatisfaction.

After thinking for a while Joe replied, "All right, you take four-sevenths of the stakes and I'll take three-sevenths."

What would $\frac{4}{7}$ of $20 be?

See Sec. A.6

---

* This dialogue reveals a famous problem in the history of probability called the *problem of the points*. This problem was first posed to the French mathematician, Blaise Pascal, in 1654 by a friend, who was an amateur mathematician and a professional gambler. In solving this problem, Pascal introduced the important notion that the proportion of the stakes deserved by the gamblers should depend on their respective probabilities of winning, should the game be continued at that point.

$$\frac{\$20}{7} \times 4 = \frac{20 \times 4}{7}$$

$$= \frac{80}{7}$$

$$\approx \$11.43$$

Event for Bob:
$B = \{th, ht, hh\}$
Event for Joe:
$J = \{tt\}$
$$P(B) = \frac{n(B)}{n(S)} = \frac{3}{4}$$
$$P(J) = \frac{n(J)}{n(S)} = \frac{1}{4}$$

$$\$20 \times \frac{3}{4} = \$15$$

$$\$20 \times \frac{1}{4} = \$5$$

See Sec. A.6.

Bob shook his head and wondered whether he should accept this compromise.

Clearly, Joe's settlement implies that he had accepted the fact that Bob should take back more money. This was so because Bob had won more games than Joe. Joe arrived at the compromise by dividing the $20 into seven equal parts, four of which went to Bob because he had won four of the seven games.

After a long pause, Bob took up a pencil and a piece of paper suggesting to Joe the following way of dividing the stakes:

"Suppose we were to continue flipping the penny two more times. Then one of us would have to win because one of the following would happen:

You win, You win	i.e., *tt*
You win, I win	i.e., *th*
I win, You win	i.e., *ht*
I win, I win	i.e., *hh*

"Therefore, the primitive sample space for flipping the penny two times, is $\{tt, th, ht, hh\}$. Of these four outcomes, the last three *th, ht,* and *hh* mean that I first win 5 games. On the other hand, only the first outcome *tt* means that you first win 5 games. So, my probability of winning the bet is $\frac{3}{4}$ and yours is $\frac{1}{4}$. Therefore, I should take three-fourths of the stakes, and not four-sevenths as suggested by you."

In short, Bob's suggestion means that the stakes should be divided in proportion to one's probability of winning the bet, should the game be continued.

After examining Bob's reasoning, Joe agreed that it was fair. Bob took $15 of the money and Joe took $5.

3.1 INTRODUCTION

In Chapters 1 and 2 we have developed the ability to determine the probability of an event from a primitive sample space. In this chapter, we shall use the probability of an event to define a useful concept called the *expected cost of a game.* This concept enables us to analyze mathematically some familiar games of chance such as "high-low" and "chuck-a-luck."

We choose these games mainly because they can be described in terms of equally likely outcomes, and hence satisfy the basic assumption of the probability theory that we have developed.

Furthermore, the mathematical analysis of these games helps us to understand better the reasons why a player almost always loses to the operator of these gambling games.

The notion of the expected cost of a game can also be used to make predictions about the frequency of the occurrence of an outcome or an event when a chance experiment is repeated a large number of times. In this situation, it is more appropriate to call the expected cost of a game the *expected number of an outcome or event* in a large number of trials. (Each repetition of a chance experiment is called a trial.)

Suppose, for example, we flip a coin 1000 times (1000 trials of the coin experiment). We guess that the two outcomes, heads and tails, would each occur with about the same frequency, that is, 500 times each. We do this even though we are not sure of the actual outcome on each flip of the coin. In other words, in spite of our inability to predict an individual experimental result, we are able to predict fairly accurately (by the use of the expected number) what the experimental results as a whole would be in the long run.

So, in this chapter, you will be given an opportunity to check the degree of agreement between your observations or experimental results and the results that you arrive at by mathematical reasoning.

$$P(A) = \frac{n(A)}{n(S)}$$

Expected cost of a game

Predict the frequency of an event.

Expected number of an event

Trial

We *predict* the number of heads and tails that should occur.

3.2 EXPECTED COST OF A GAME

In throwing a die, we know the primitive sample space is

$$S = \{1,2,3,4,5,6\}.$$

Suppose your friend Jack bets a dollar on the outcome 6 on one roll of a die. How much should Jack win from you if the outcome 6 does occur?

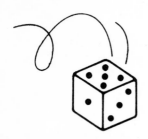

Is this bet fair to Jack?

Let us say you pay Jack the same amount as his bet, that is, a dollar in this case. What advantage do you now have over Jack? The advantage is that your probability of winning is $\frac{4}{6}$ higher than Jack's for the same amount of money. Why?

Since Jack has only one way to win the $2 (that is, the outcome 6), then the probability of Jack's winning is

$$\frac{1}{6} \approx 17\%.$$

Changing a Fraction to a percent. See Sec. C.2.

On the other hand, you have five outcomes, namely, 1,2,3,4,and 5, favorable to your winning the $2. Your probability of winning is

$$\frac{5}{6} \approx 83\%.$$

$$\frac{5}{6} - \frac{1}{6} = \frac{5-1}{6}$$
$$= \frac{4}{6}$$
$$= \frac{2}{3}$$
$$\approx 67\%$$

Comparing these two probabilities of winning, it is clear that you would win most of the time. To be precise, your probability of winning is

$$\frac{5}{6} - \frac{1}{6} = \frac{4}{6} \approx 67\%$$

higher than Jack's. Thus, such a bet is very unfair to Jack.

The bet could be made fair by giving the bettors an equal probability of winning.

At this point, you may suggest that to make the bet fair the probability of winning should be the same for both of you, since each of you pays the same amount of money, $1, to gamble. That is, each of you should have a 50% probability of winning the money. The 50% probability would mean that Jack should have three outcomes of the six favorable to his winning, and the remaining three outcomes should be in your favor. Such a deal is clearly fair to both of you. However, there is another more interesting way of making this bet fair.

Another way of making the bet a fair one.

Suppose we would like to maintain the conditions that Jack has only one favorable outcome, and you have five favorable outcomes. These conditions may seem to suggest that you should pay $5 to gamble since you have five favorable outcomes, whereas Jack should have to pay $1 for one favorable outcome. In other words, you should lose $5 if the die comes up 6, and win $1 if the die comes up 1,2,3,4, or 5.

Such a deal is equivalent to saying that the amount of money that each of you pays to gamble should be *in proportion to* the number of outcomes favorable to your winning. Another way of saying this is that *the amount of money that a person pays to gamble should be proportional to his probability of winning*, since a greater number of favorable outcomes means a higher probability of winning. In order to make the die-throwing bet fair we are proposing the following:

If Jack bets $1 and his probability of winning is $\frac{1}{6}$, then you should have to bet $5 if your probability of winning is $\frac{5}{6}$.

The $5 you bet is called your *expected cost of the game*. The sum of $6 is called the *prize of winning*.

Definition. *A person's* expected cost *of a game is equal to the product of the probability of the event in favor of his winning, and the amount he collects if the event occurs.*

Suppose we let E stand for a person's expected cost of a game, and $P(A)$ the probability of the event A in favor of his winning, and M the amount he collects if the event occurs. Then the definition can be written as:

$$E = P(A) \times M$$

Now, let's look at the problem we have been discussing. Jack bets $1 and you bet $5. He wins if the die comes up 6 and you win if it comes up anything else. So, $M = \$6$.

If Jack's favorable event is $A = \{6\}$, then $P(A) = \frac{1}{6}$, and Jack's expected cost of the game is

$$E = P(A) \times M = \frac{1}{6} \times 6 = \$1.$$

in proportion to

You must bet *five times* as much since your chance of winning is *five times* as great.

Expected cost of a game.

$E = P(A) \times M$
  or
$E = M \times P(A)$
  why?

The winner collects $6.

M(prize) = $6.

Reciprocals

The reciprocal of a number is 1 divided by the number.

Examples:

1. The reciprocal of 6 is $\frac{1}{6}$

2. The reciprocal of 7 is $\frac{1}{7}$

3. The reciprocal of $\frac{1}{6}$ is 6

4. The reciprocal of $\frac{1}{7}$ is 7.

See Sec. A.7.

Fair or not fair.

Similarly, if your favorable event is $B = \{1,2,3,4,5\}$, then $P(B) = \frac{5}{6}$, and your expected cost of the game is

$$E = P(B) \times M = \frac{5}{6} \times 6 = \$5.$$

This is a fair bet for you since you put up $5 and your expected cost is $5. It is also fair for Jack who bets $1 and has an expected cost of $1.

If your expected cost of a game comes out to be the same as your bet, then we say that the game is fair to you. If it comes out to be less than your bet, then we say that the game is unfair to you.

Questions:

1. What can we say if the expected cost of the game is more than your bet?

2. Is the game fair to the other player?

Examples:

1. A person wins $8 if he draws an ace from a set of 10 different cards from ace to 10. How much should he pay for one draw?

Solution:

$S = \{ace, 2, 3, 4, \ldots, 10\}$
$n(S) = 10$

The amount to be collected $M = \$8$. There is only one ace out of 10 cards, so the favorable event is $A = \{ace\}$ and thus the probability of drawing an ace is $P(A) = \frac{1}{10}$. Therefore, the expected cost

$$E = \frac{1}{10} \times 8$$

$$= \frac{8}{10}$$

$$= \$.80$$

Hence, the game is fair if he pays 80¢ for each draw of a card.

Note that in this case the person collects $8 if he wins. This includes the $.80 that he bet. So the dealer loses $7.20. This is important to note because the dealer's expected cost is

$$E = \frac{9}{10} \times 8 = \$7.20$$

Suppose the player pays a dollar for one draw. Then, he pays 20¢ extra. This means that he is expected to lose 20¢ for each draw. In other words, if he plays this game a large number of times, he will lose at the rate of 20¢ per game.

2. In a lottery, 10,000 tickets are sold. If the grand prize is $2,000, $2\ what is the fair price for a person to pay for a ticket?

Solution:

We regard each number of a lottery ticket as a possible outcome. There are 10,000 tickets sold and hence the sample space for the experiment of drawing a ticket has 10,000 possible outcomes. If a person buys one ticket, then his favorable event A has only one outcome, that is, the number of the lottery ticket that he bought. So, we have $P(A) = \frac{1}{10000}$.

Multiplication of two fractions

1. We multiply their numerators to obtain the numerator of the product.

2. We multiply their denominators to obtain the denominator of the product.

Examples:

1. $\frac{4}{7} \times \frac{2}{5} = \frac{4 \times 2}{7 \times 5}$

$$= \frac{8}{35}$$

2. $\frac{3}{11} \times \frac{2}{9} = \frac{\overset{1}{\cancel{3}}}{11} \times \frac{2}{\underset{1}{\cancel{3} \times 3,}}$ why?

$$= \frac{1 \times 2}{11 \times 3}$$

$$= \frac{2}{33}$$

Since a whole number can be represented as a fraction with 1 as a denominator, we can regard the multiplication of a whole number by a fraction as the multiplication of two fractions.

See Sec. A.6, B.1

$S = \{0001, 0002, 0003, \ldots, 10{,}000\}$
$n(S) = 10{,}000$

$$\frac{2000}{10,000}$$

$$= \frac{1 \times 2 \times 1000}{5 \times 2 \times 1000}$$

$$= \frac{1}{5} \times \frac{2}{2} \times \frac{1000}{1000}$$

$$= \frac{1}{5}$$

$$S = \left\{ 001, 002, 003, \dots, 1000 \right\}$$
$$n(S) = 1,000$$

Since $M = \$2,000$, then the expected cost for a ticket is

$$E = \frac{1}{1\,0000} \times 2000$$

$$= \frac{2000}{1\,0000}$$

$$= \frac{2}{10}$$

$$= \frac{1}{5}$$

$$= \$.20$$

This means that the fair price or expected cost is 20 cents per ticket. Note that in this example the lucky ticket pays $2,000 and the remaining 9,999 tickets will not pay anything at all.

3. A person pays 50¢ for one of 1,000 raffle tickets for a prize worth $450. Does he have a fair deal?

Solution:

Since the probability of winning for each ticket is $\frac{1}{1000}$ and

$M = \$450$, then his expected cost

$$E = \frac{1}{1000} \times 450$$

$$= \$.45$$

This means that the fair price is 45¢ per ticket. Since he pays 50¢ for a ticket, he then has paid 5¢ more than the fair price (his expected cost). He does not have a fair deal.

4. If the person in Example 3 buys 10 tickets, what should he pay?

Solution:

We have been regarding each ticket of the 1000 as a possible outcome, and we thus have 1,000 possible outcomes. Since the person buys 10 tickets, he thus has 10 possible outcomes favorable to his winning of the prize. So, we have $n(A) = 10$ tickets,

$P(A)$, $= \frac{10}{1000}$, and his expected cost for 10 tickets is

$$E = \frac{10}{1000} \times 450$$
$$= \$4.50.$$

$4.50 is the fair price for 10 tickets.

We may also find the fair price for 10 tickets by using the fact obtained in Example 3 that the fair price for each ticket is 45¢. Hence, the fair price for 10 tickets is $.45 \times 10 = \$4.50$. Similarly, to find the fair price for 100 tickets we multiply 45 cents by 100, which is $45.00.

5. A player pays a dollar to throw a die once. He wins two dollars (including his one dollar) if he throws a number larger than the number thrown by the dealer; otherwise, he loses his dollar.

    a)   Is this game fair to the player?

    b)   How much is he expected to lose in a game?

Solution:

a) Before we can calculate the probability of the player's winning, we first have to figure out the primitive sample space for this game.

Each game consists of *throwing a die twice*: once by the dealer, and once by the player. That is, the outcome of this game is a pair of numbers, each runs from 1 through 6. So, the primitive sample space is the same as that for throwing *two dice once*:

$$S = \begin{Bmatrix} (1,1), & (1,2), & (1,3), & (1,4), & (1,5), & (1,6) \\ (2,1), & (2,2), & (2,3), & (2,4), & (2,5), & (2,6) \\ (3,1), & (3,2), & (3,3), & (3,4), & (3,5), & (3,6) \\ (4,1), & (4,2), & (4,3), & (4,4), & (4,5), & (4,6) \\ (5,1), & (5,2), & (5,3), & (5,4), & (5,5), & (5,6) \\ (6.1), & (6,2), & (6,3), & (6,4), & (6,5), & (6,6) \end{Bmatrix}$$

The first numeral of each outcome denotes the number thrown by the dealer, and the second by the player.

How many of these outcomes have the first number less than the second?

Multiplication of two decimals:

1. Ignore the decimal points, and multiply the two numbers as if they were whole numbers.
2. Insert the decimal point in the product so that the number of decimal places in the product is the sum of the numbers of decimal places in the given two numbers.

Examples

1. $12.5 \times 1.3 = 16.25$

$$\begin{array}{r} 12.5 \\ \times\ 1.3 \\ \hline 375 \\ 125 \\ \hline 16.25 \end{array}$$

2. $.45 \times 100 = 45.0$

$$\begin{array}{r} .45 \\ \times\ 100 \\ \hline 45.00 \end{array}$$

See Sec. B.6

Look at the pairs of numbers in the triangle. Is the second number always greater than the first?

Going over the 36 outcomes in S, we see that there are 15 outcomes with the first number less than the second.

So the probability of the player's winning is $\frac{15}{36}$, that is $P(A) = \frac{15}{36}$, where $n(A) = 15$.

Since $M = \$2$, we have

$$E = \frac{15}{36} \times 2$$
$$= \frac{5}{6}$$
$$\approx .83$$

This means that he should pay about 83¢ for a game. Since we know that he pays a dollar, he therefore pays about 17¢ extra. Therefore, the game is unfair to the player.

Subtraction of decimals

1. Write one number above the other with their decimal points aligned vertically.
2. Subtract as if the numbers were whole numbers.
3. Place the decimal point in the difference in vertical alignment with the other decimal points.

Example

$$
\begin{array}{r}
1.00 \\
-0.83 \\
\hline
.17
\end{array}
$$

b) We have calculated that his expected cost of the game is 83¢. He is thus expected to lose

$$\$1. - \$.83 = \$.17,$$

or 17¢ per game in the long run.

EXERCISES 3.2

1. Thirteen playing cards, ace through king, are placed randomly with faces down on a table. The prize for guessing correctly the value of any given card is $1. What would be a fair price to pay for a guess?

Method:
1. Find $n(S)$
2. Find A and $P(A)$
3. Identify M
4. Use formula $E = P(A) \times M$

2. A man gets 10 dollars if he throws a double on a single throw of a pair of dice. How much should he pay for a throw?

$n(S) = 36$

3. A player bets a dollar on a two-digit number. He wins $75 if he draws his number from the set of all two-digit numbers, $\{00, 01, 02, ..., 99\}$; otherwise, he loses his dollar.

   a)   Is this game fair to the player?

   b)   How much is he expected to lose in a game?

$n(S) = 100$

4. A player bets a dollar on a three-digit number. He wins $600 if he draws his number from the set of all three-digit numbers, $\{000, 001, 002,..., 999\}$; otherwise, he loses his dollar.

   a)   Is this game fair to the player?

   b)   How much is the player expected to lose in a game?

   c)   Which of the two games described in this and the last questions is a better game for the player?

$n(S) = 1,000$

5. In a lottery, 1,000 tickets are sold at 20¢ each. If a prize worth $150 is to be paid for the lucky ticket, do these lottery tickets cost too much?

6. In a raffle, 1,000 tickets were sold at 10¢ each or 10 for 90¢. If the prize is worth $90, is the raffle fair or unfair to the ticket buyer?

There are two questions in this problem.

In this example, expected cost is called the *premium*.

7. According to an insurance mortality table, the probability that a person in the 20-25 age group will die within a year is 0.002. If a person in this age group wants a policy that pays his beneficiary $10,000 should he die within a year, what is the basic premium that he would be expected to pay?

8. How much should a player pay for a chance to play the game of throwing two dice, if he receives $5 when he throws a sum of:

   a)   5 or 6 on the two dice?

   b)   5 or less on the two dice?

   c)   5 or more on the two dice?

   d)   not 5 on the two dice?

9. *High-low (Over-under)*. The following diagram illustrates a gambling game known as High-low. It is played with two dice.

Under	Seven	Over
2, 3, 4,	7	8, 9, 10,
5, 6		11, 12

A player bets a dollar on one of the three events:

a) Under; which includes all outcomes of the two dice that give a sum of 2, 3, 4, 5, or 6.

b) Seven; which includes all outcomes of the two dice that give a sum of 7.

c) Over; which includes all outcomes of the two dice that give a sum of 8, 9, 10, 11, or 12.

If the event Under occurs, the player who bets on this event wins $2. The same thing is true for the event Over. However, if the event Seven occurs, the player who bets on it wins $5.

a) Is this game fair to the player?

b) Which of the three events allows the player to lose the least?

10. Roulette (See Exercise 9, Review Exercises for Chapter I). A player bets a dollar on the color red (or black). He will win two dollars if the ball finally rests on one of the 18 red (or black) compartments: otherwise, he loses his dollar.

a) Is this game fair to the player?

b) How much is the player expected to lose per game?

3.3 MORE PROBLEMS ON EXPECTED COST

In a lottery, there is often more than one prize. For example, let us change Example 2 in Section 3.2 so that there is a second prize worth $1,000 and a third prize worth $500. In this case, what is the fair price to pay for a ticket?

Since the probability that a ticket will win the first prize of $2000 is $\frac{1}{10,000}$, then the fair price to pay for a ticket to win the first prize is $\frac{1}{10,000} \times 2000 = .20$; similarly, the fair price to pay for a ticket to win the second prize is $\frac{1}{10,000} \times 1000 = .10$, and that for the third prize is $\frac{1}{10,000} \times 500 = .05$.

Since a ticket can win each of the 3 prizes, the fair price to pay for a ticket is the sum of the 3 fair prices:

$$.20 + .10 + .05 = .35,$$

We see that

$$\$.35 \times 10,000 = \$3,500$$

This amount of money equals in fact the sum of all the 3 prizes. In other words, a fair price for a ticket can be found as follows:

$$(\frac{1}{10,000} \times 2000) + (\frac{1}{10,000} \times 1000) + (\frac{1}{10,000} \times 500)$$

$$= \frac{1}{10,000} \times (2,000 + 1,000 + 500)$$

$$= \frac{1}{10,000} \times 3,500$$

$$= \$.35$$

Examples:

1. In a raffle, 1,000 tickets are sold. The first prize is $450, the second prize $300, the third prize $200, and the fourth prize $50. What is the fair price to pay for a ticket?

Assume that the same ticket can win all three prizes.

Fair price for the 1st prize = 20¢.

Fair price for a ticket:
```
 .20
 .10
+.05
─────
$.35
```

Distributive Law:
Since each of the three products has $\frac{1}{10,000}$ as a factor, we can "factor it out."

Assume that the same ticket can win all four prizes.

Solution:

The fair price to pay for a ticket

$$= (\frac{1}{1000} \times 450) + (\frac{1}{1000} \times 300) + (\frac{1}{1000} \times 200) + (\frac{1}{1000} \times 50)$$

Distributive Law:

Here $\frac{1}{1000}$ is a common factor.

$$= \frac{1}{1000} \times (450 + 300 + 200 + 50)$$

$$= \frac{1}{1000} \times 1000$$

$$= \$1$$

2. A box contains 5 one-dollar bills, and 5 five-dollar bills. You are to reach into the box and draw one bill, which you may then keep. What is the fair price to pay to have one draw?

Solution:

$$S = \{o_1, o_2, o_3, o_4, o_5, f_1, f_2, f_3, f_4, f_5\}$$

$$n(S) = 10$$

$$P(\{o_1, o_2, o_3, o_4, o_5\}) = \frac{1}{2}$$

$$P(\{f_1, f_2, f_3, f_4, f_5\}) = \frac{1}{2}$$

We may regard the first prize worth \$5, and the second prize worth \$1. Since the probability of drawing either of the 2 kinds of bills is $\frac{5}{10} = \frac{1}{2}$, the fair price to pay for one draw

$$= (\frac{1}{2} \times 5) + (\frac{1}{2} \times 1)$$

$$= \frac{1}{2} \times (5 + 1)$$

$$= \frac{1}{2} \times 6$$

$$= \$3$$

Expected cost = Fair price.

3. In the last example, what is the expected cost if there are only 3 five-dollar bills, and 5 one-dollar bills?

Solution:

Since there are 8 bills altogether in the box of which 3 are five-dollar bills, then the probability of drawing a five-dollar bill is $\frac{3}{8}$. The probability of drawing a one-dollar bill is $\frac{5}{8}$. Why? The expected cost is

$$E = (\frac{3}{8} \times 5) + (\frac{5}{8} \times 1)$$

$$= \frac{15}{8} + \frac{5}{8}$$

$$= \frac{15+5}{8}$$

$$= \frac{20}{8}$$

$$= \$2.50$$

$$S = \{o_1,o_2,o_3,o_4,o_5,f_1,f_2,f_3\}$$

$$n(S) = 8$$

$$P(\{f_1,f_2,f_3\}) = \frac{3}{8}$$

$$P(\{o_1,o_2,o_3,o_4,o_5\}) = \frac{5}{8}$$

This means that you should pay $2.50 for a chance to draw the bill.

4. In Example 2, what is your expected cost if there are 2 ten-dollar bills, 3 five-dollar bills, and 5 one-dollar bills.

Solution:

Since there are 10 bills altogether, of which 2 are ten-dollar bills, then the probability of drawing a ten-dollar bill is $\frac{2}{10}$. The probability of drawing a five-dollar bill is $\frac{3}{10}$, and the probability of drawing a one-dollar bill is $\frac{5}{10}$. Why? The expected cost is

$$E = (\frac{2}{10} \times 10) + (\frac{3}{10} \times 5) + (\frac{5}{10} \times 1)$$

$$= \frac{20+15+5}{10}$$

$$= \frac{40}{10}$$

$$= \$4$$

$$S = \{t_1,t_2,f_1,f_2,f_3,o_1,o_2,o_3,o_4,o_5\}$$

$$n(S) = 10$$

$$P(\{t_1,t_2\}) = \frac{2}{10}$$

$$P(\{f_1,f_2,f_3\}) = \frac{3}{10}$$

$$P(\{o_1,o_2,o_3,o_4,o_5\}) = \frac{5}{10}$$

This means that you should pay $4 for a chance to draw a bill.

In summing up the last four examples, it may be concluded that:

If P_1 is the probability of winning the amount M_1, P_2 the probability of winning the amount M_2, and P_3 the probability of winning the amount M_3, then the expected cost is

$$E = M_1 P_1 + M_2 P_2 + M_3 P_3$$

Can you identify P_1, P_2, P_3, M_1, M_2, and M_3 in this example 4?

More Examples:

5. You pay $.05 to flip a coin twice. You receive $1 for two heads, $.50 for one head, and nothing for no heads. Have you paid too much in this game?

Solution:

In this problem,

$$S = \{hh, ht, th, tt\}$$
$$n(S) = 4$$
$$P(\{hh\}) = \frac{1}{4} = P_1$$
$$P(\{ht, th\}) = \frac{1}{2} = P_2$$
$$P(\{tt\}) = \frac{1}{4} = P_3$$

$$P_1 = \frac{1}{4}, \quad M_1 = \$1$$
$$P_2 = \frac{1}{2}, \quad M_2 = \$.50$$
$$P_3 = \frac{1}{4}, \quad M_3 = 0$$

Hence, the expected cost is:

$$E = (1 \times \frac{1}{4}) + (.50 \times \frac{1}{2}) + (0 \times \frac{1}{4})$$
$$= \frac{1}{4} + \frac{.50}{2}$$
$$= \frac{1}{4} + \frac{1}{4}$$
$$= \frac{2}{4}$$
$$= \frac{1}{2}$$
$$= \$.50$$

This means that you have paid a fair amount to play.

6. *Chuck-a-luck (Big Six):* A player bets a dollar on one of the numbers 1 through 6. Assume that the player bets on the number 6. *Three* dice are then rolled. The player wins $1 from the dealer if the number 6 appears on one of the dice; $2 if it appears on

two of them; and $3 if it appears on all three dice. The player loses his dollar if none of the dice has a 6.

a) Is this game fair to the player?

b) How much is the player expected to lose per game?

The dealer pays:
$0 for no 6
$1 for one 6
$2 for two 6's
$3 for three 6's

The player gets:
(including his own $1)
$2 for one 6 ($M_3$)
$3 for two 6's ($M_2$)
$4 for three 6's ($M_1$)

Solution:

From Example 3 of Chapter 2 on page 86, we know the probability of getting three 6's is $\frac{1}{216}$, which we may regard as the probability of winning the first prize of $4. Thus,

$$P_1 = \frac{1}{216} , \ M_1 = \$4$$

$$P_2 = \frac{15}{216} , \ M_2 = \$3$$

$$P_3 = \frac{75}{216} , \ M_3 = \$2$$

Therefore, the money that the player is expected to pay is

$$E = (\frac{1}{216} \times 4) + (\frac{15}{216} \times 3) + (\frac{75}{216} \times 2)$$

$$= \frac{4}{216} + \frac{45}{216} + \frac{150}{216}$$

$$= \frac{199}{216}$$

$$\approx .921$$

$$\approx \$.92$$

See Sec. A.5, A.6

See Sec. B.3, B.4

a) Since the player pays a dollar to play one game, he pays more than the expected cost. Thus, the game is unfair to the player.

b) Since he pays 8¢ extra for a game, he is expected to lose 8¢ per game. (Play this game in class and determine whether you, the player, lose at the rate of 8¢ per game.)

A note on the idea of odds:

In games of chance, such statements are sometimes used:

1. "The *odds in favor* of getting a 6 on one roll of a die are 1 to 5."

Odds in favor

2. "The *odds against* drawing an ace from a deck of 52 cards are 12 to 1."

Odds against

What do the phrases "odds in favor of" and "odds against" really mean?

Mathematically, they are defined as follows:

$$\text{Odds in favor of an event } A = \frac{P(A)}{P(\overline{A})}$$

$$\text{Odds against an event } A = \frac{P(\overline{A})}{P(A)}$$

$\overline{A}$ = Complement of A

In (1), we are given the event $A = \{6\}$. Thus, we have

$$\overline{A} = \{\text{not } 6\}$$

$$P(A) = \frac{1}{6}$$

$$P(\overline{A}) = \frac{5}{6}$$

Hence, the odds in favor of the event $\{6\}$

$$= \frac{\frac{1}{6}}{\frac{5}{6}}$$

$$= \frac{1}{5} \text{ or 1 to 5.}$$

The odds against the event $\{6\}$

$$= \frac{\frac{5}{6}}{\frac{1}{6}}$$

$$= \frac{5}{1} \text{ or 5 to 1.}$$

Check:

$P(\overline{A}) = 1 - P(A)$

$\frac{5}{6} = 1 - \frac{1}{6}$

$\frac{\frac{1}{6}}{\frac{5}{6}} = \frac{\frac{1}{6} \times \frac{6}{5}}{\frac{5}{6} \times \frac{6}{5}}$, Why?

$= \frac{\frac{1}{6} \times \frac{6}{5}}{1}$

$= \frac{1}{6} \times \frac{6}{5}$

$= \frac{1}{5}$

Note that the fraction that expresses the odds *in favor* of the event A is the *reciprocal* of the fraction that expresses the odds *against* the event A.

reciprocal

In (2), $A = \{\text{an ace}\}$. Thus, we have

$$\overline{A} = \{\text{not an ace}\}$$

$$P(A) = \frac{4}{52} = \frac{1}{13}$$

$$P(\overline{A}) = \frac{48}{52} = \frac{12}{13}$$

Division of Fractions
When dividing two fractions, take the fraction which is the denominator (divisor), invert it, and then multiply by the fraction which is the numerator (dividend).

See Sec. A.7

Hence, the odds *in favor of* drawing an ace

$$= \frac{\frac{1}{13}}{\frac{12}{13}}$$

$$\frac{\frac{1}{13}}{\frac{12}{13}} = \frac{1}{13} \times \frac{13}{12} \text{, Why?}$$

$$= \frac{1}{12} \text{ or 1 to 12}$$

$$= \frac{1}{12}$$

The odds *against* drawing an ace

$$= \frac{\frac{12}{13}}{\frac{1}{13}}$$

$$= \frac{12}{1} \text{ or 12 to 1.}$$

Do you see how odds and probabilities are related?

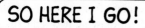

EXERCISES 3.3

Method
1. Find $n(S)$
2. Find P_1 and M_1
3. Find P_2 and M_2
4. Find P_3 and M_3
5. Use formula:
   $E = P_1 M_1 + P_2 M_2 + P_3 M_3$

1. A bureau has four drawers. One contains a ten-dollar bill, one a five-dollar bill, one a one-dollar bill and the fourth drawer is empty. A man is to open one of these drawers and to keep its contents. How much should he be expected to pay to open a drawer?

What is the probability of getting 2 heads, 1 head, 0 heads?

2. You pay a dollar to toss two coins. If you toss two heads, you get two dollars (including your one dollar); if you toss only one head, you get back your dollar; and if you toss no heads, you lose your one dollar. Is this a fair game to play?

3. In a raffle, 1,000 tickets are sold at 50¢ each. The first prize is $100, and there are three second prizes of $50 each. How much should you pay for a ticket?

4. In a raffle, 1,000 tickets are sold at 50¢ each. The first prize is $100. There are two second prizes of $50 each, and 5 third prizes of $10 each (there are 8 prizes in all). Bob buys one ticket. How much should he pay?

$n(S) = 36$

5. You toss two dice. You win 2 dollars from Bill if one of the dice comes up 6. If the sum on the two dice is 4 or less, you win a dollar. How much should you pay Bill to toss two dice once?

6. You roll two dice once. You receive $1 for a 6, and $2 for two 6's. What is your expected cost in this game?

$n(S) = 216$

7. Three coins are tossed. You win $1.50 for 3 heads, $1.00 for two heads, and $.50 for one head. How much should you pay to play one game?

$n(S) = 52 \times 51$
$\quad\quad\ = 2652$

8. In a game, you draw two cards at random *without* replacement from a deck of 52 playing cards. You win a dollar for one ace and two dollars for two aces. What is a fair price for you to pay to play one game?

9. You draw two balls in succession *with* replacement from a bag containing 9 balls (1 red, 3 green, and 5 blue). You win $5 for drawing two red balls, and $3 for a red and a green ball. What is a fair price to pay to play one game?

$n(S) = 21$, why?
$P_1 = \dfrac{1}{21}; M_1 = 10$

$P_2 = \dfrac{10}{21}; M_2 = 6$

$P_3 = \dfrac{10}{21}; M_3 = 2$

10. Joe is to draw *two* bills together from a bag that contains 5 one-dollar bills and 2 five-dollar bills. How much should Joe pay for one draw?

11. In the two-coin experiment,

   a)   What are the odds in favor of flipping a head and a tail?

   b)   What are the odds against flipping a head and a tail?

   c)   What are the odds in favor of flipping 2 heads or 2 tails?

   d)   What are the odds against flipping 2 heads or 2 tails?

12. In the die experiment,

   a)   What are the odds in favor of rolling a number less than 5?

   b)   What are the odds against rolling a number less than 5?

   c)   What are the odds in favor of rolling an even number?

   d)   What are the odds against rolling an even number?

3.4 EXPECTED NUMBER OF AN EVENT

Suppose that you and Jack have agreed to play the game of throwing a die on the conditions that we have discussed in Section 3.2. That is, Jack pays $1 to bet on the outcome 6, and you pay $5 to bet on the other five outcomes, 1,2,3,4, and 5. Both of you have also decided to play 30 times. Who would you guess would come out ahead after the 30 games?

   Of course, you would not bet with Jack if you were certain that you would lose. You realize that whether you win or lose depends on the number of times that the die comes up 6. The important question is: how many times would you *expect* the die to come up 6 in 30 throws?

What would you expect?

$P(\{6\}) = \dfrac{1}{6}$

$\approx 17\%$

This means that the bet is "fair."

The more you roll the die, the closer you will come to getting the outcome 6 one-sixth of the time.

Expected number of an event.

$E = P(A) \times n$
or
$E = n \times P(A)$

We know that the probability of throwing a 6 is $\dfrac{1}{6} \approx 17\%$. That is, if we throw a die a large number of times, we would expect to get the outcome 6 about 17% of the time (or, to be more exact, $\dfrac{1}{6}$ of the time). Now, we know that the total number of throws is 30, and thus the expected number is 17% of 30, or to be exact

$$30 \times \frac{1}{6} = 5 \text{ times.}$$

From this reasoning, you expect to get the outcome 6 five times out of 30 throws. If this happens, then you would lose $25 to Jack. However, you would also have won $25 from Jack for the other 25 throws which did not result in the outcome 6. *This means that you and Jack would break even after 30 games if the expected number of times for the outcome 6 comes true.*

In practice, we know that it is not likely that you will get exactly five 6's in 30 throws. You will probably get between three and seven 6's. (Take a die and throw it 30 times to verify this.) However, if you throw a die a *large number of times*, say 3,000 times, then you would discover that the number of 6's that you get is very close to 500 times or $\dfrac{1}{6}$ of the 3,000 throws. (Use your experimental results in the die experiment which you did in Exercises 1.4 to verify this prediction.)

Since you and Jack have only 30 throws of a die, neither of you can *really* predict who will win after the 30 games. It all depends on your *luck*, or on *chance*

Definition. *The expected number of times that an event A occurs is equal to the product of the probability of the event A and the total number of trials.*

Suppose we let *E* stand for the expected number of times that the event *A* occurs, and let *n* stand for the total number of trials. Then the definition can be written as:

$$E = P(A) \times n$$

Examples:

1. *Coin-die Experiment: Throwing a coin and a die together.* If this experiment is repeated 72 times, then what is the expected number of times that each of the following events will occur:

   a) $A = \{h6\}$?

   b) $B = \{h6, t6\}$?

   c) $C = \{h1, h2, h3, h4, h5, h6\}$?

$n(S) = 12$

See p. 53

Solution:

   a) The number of trials $n = 72$. Since $P(A) = \frac{1}{6}$, then the expected number of times that the event A will occur is

$$E = \frac{1}{12} \times 72$$
$$= 6$$

This means that we expect to see the event A or the outcome h6 occur about 6 times out of 72 trials.

$E = \frac{1}{12} \times 72$
$= \frac{72}{12}$
$= 6$

See Sec. A.6.

   b) Since $P(B) = \frac{2}{12}$, then the expected number of times that the event B will occur is

$$E = \frac{1}{6} \times 72$$
$$= 12$$

This means that we expect to see the outcomes t6 and h6 occurring a total of about 12 times out of 72 trials.

   c) Since $P(C) = \frac{6}{12} = \frac{1}{2}$ then the expected number of times that the event C will occur is

$$E = \frac{1}{2} \times 72$$
$$= 36.$$

Explain this result.

2. It has been found that students who register for a certain history course have a probability of .80 to attend the course. How many students would you expect to attend the course if there are 40 students registered for the course?

Solution:

We regard the total number of registered students as the total number of trials. Thus $n = 40$, and $P(A) = .80$, where A is the event consisting of a registered student who will attend the course. Therefore, the expected number of registered students who will attend the course is

$$E = .80 \times 40 = 32$$

This means that about 32 registered students will attend the history course.

```
  40
X .80
------
32.00
```

See Sec. B.6.

© 1968 United Feature Syndicate, Inc.

EXERCISES 3.4

1. A card is drawn randomly from a deck of 52 cards *100 times*. (The drawn card is put back into the deck before the next card is drawn). Find the number of times that each of the following events is expected to occur.

First find the probability for each event.

   a) $A = \{$an ace$\}$
   b) $B = \{$a spade$\}$
   c) $C = \{$a picture card$\}$
   d) $D = \{$a red card$\}$
   e) $E = \{$not a heart$\}$

If not a heart, what else can it be?

2. Two dice are rolled 60 times. Find the expected number for each of the following events:

Find the probability for each event first.

   a) $A = \{$a sum of 7$\}$
   b) $B = \{$a double$\}$

(2,2) is a double.

   c) $C = \{$not a double$\}$
   d) $D = \{$an even sum$\}$

(2,4) gives an even sum of 6

3. A certain kind of light bulb has been found to have .02 probability of being defective. A businessman receives 500 light bulbs of this kind. How many of these bulbs would he expect to be defective?

4. A student enrolled in a math course has 0.9 probability of passing the course. In a class of 20 students, how many would you expect to fail the math course?

5. A doctor has found that the probability that a patient who was given a certain drug would have unfavorable reactions to the drug was 0.002. If a group of 500 patients are going to be given the drug, how many of them would the doctor expect to have unfavorable reactions?

SUMMARY

Fair Game: A game of chance is said to be *fair* if a player pays the same amount as his expected cost of the game. Otherwise the game of chance is said to be *unfair* to the player or the dealer.

Terms and Symbols:

Terms	Symbols	Page
Expected cost of a game.	E	102
Expected number of an event.	E	119
Odds		115

Formulas:

1. $E = P(A) \times M$

   Where E is the expected cost of a game, and $P(A)$ is the probability of the event A favorable to the winning of the prize M.

2. $E = P(A) \times n$

   Where E is the expected number of an event A, which has the probability $P(A)$ of occurring in each of the n trials.

3. $E = P_1 M_1 + P_2 M_2 + P_3 M_3$

   Where P_1 is the probability of winning the prize M_1; P_2, the prize M_2; and P_3, the prize M_3.

4. Odds in favor of the event $A = \dfrac{P(A)}{P(\bar{A})}$

5. Odds against the event $A = \dfrac{P(\bar{A})}{P(A)}$

REVIEW EXERCISES

1. A European roulette wheel has only thirty-seven compartments, eighteen red, eighteen black, and one green. A player will be paid two dollars (including his one-dollar bet) if he picks correctly the color of the compartment in which the ball finally rests. Otherwise, he loses a dollar. Is the game fair to the player? Explain.

2. In a lottery, 1000 tickets are sold at 25 cents each. There are three cash prizes, $100, $50, and $30. Alice buys five tickets.

    a)   What would have been a fair price for a ticket?

    b)   How much extra did Alice pay?

3. Bob draws a card from a deck of 52 cards. He receives 40¢ for a heart, 50¢ for an ace, and 90¢ for the ace of hearts. If the cost of a draw is 15¢, should he play the game? Explain.

4. In a certain game, a player has the probability $\frac{1}{7}$ of winning a prize worth $89.99, and the probability $\frac{1}{3}$ of winning another prize worth $49.99. What is the expected cost of the game for the player?

5. Frank pays 70¢ to play a certain game. He draws *two* balls (together) from a bag containing two red balls and 4 green balls. He receives a dollar for each red ball that he draws. If he draws no red balls, he loses his 70¢. Has he paid too much? By how much?

6. A card is drawn from a deck of 52 cards.

    a)   What are the odds in favor of drawing a heart?

    b)   What are the odds against drawing a heart?

    c)   What are the odds in favor of drawing a red card?

    d)   What are the odds against drawing a red card?

7. In the two-die experiment,

    a)   What are the odds in favor of rolling a sum of 7?

    b)   What are the odds against rolling a sum of 7?

    c)   What are the odds in favor of rolling a sum less than 7?

    d)   What are the odds against rolling a sum less than 7?

8. Thirty dice are thrown.

    a)   What is the expected number of 1's?

    b)   How many dice must be rolled if the expected number of 1's is to be 10?

9. The probability that a new television set is defective is 0.015. In a shipment of 1000 television sets, how many of them would you expect to be defective?

10. A typist has the probability .008 of making at least one typographical error on a page. How many pages in a book of 432 pages would you expect to contain at least one typographical error?

Relative Frequency 4

4.1 INTRODUCTION

In the first three chapters, we found the probability of an event on the assumption that all outcomes of an experiment are equally likely to occur. We accepted the assumption, either because we observed that the relative frequencies of the outcomes were about the same, or because we had no reason to expect that any outcome was more likely to occur than another.

In this chapter, we will deal with experiments and situations whose outcomes seem to challenge the equally likely assumption. One example would be an experiment with a biased or "loaded" die. Since the die is known to be biased, then the six outcomes cannot be assumed to have equal probability. Their unequal probabilities can be readily approximated by their relative frequencies in a large number of throws of the die. Another example would be the life span of people. We have observed that very few people live to the age of 90; but a very large number of people survive the first 30 years of their lives. This means, among other things, that a person of age 20 has a high probability of living to 30, but a small probability of living to 90. These probabilities, which are of great interest to life insurance companies, can be determined or approximated by relative frequencies obtained from an actual examination of samples of the population.

Because of the importance of the *relative frequency as an approximation for the probability*, we will discuss this concept in more detail. We will also study the notion of averages, in particular *the mean,* and use it to increase the accuracy of the approximation of the probability by relative frequency. An accurate approximation will allow us to better understand the situations and experiments under study, and hence draw better and more correct conclusions about them.

Conclusions are often more readily drawn and understood if the collected data are tabulated and graphed. So, as an example, the use

Outcomes not all equally likely to occur.

Life expectancy

of the *frequency distribution table* and its *bar graph* will be discussed in this chapter.

4.2 FREQUENCY DISTRIBUTION

When collecting numerical data from a number of trials of an experiment, it is essential that we collect the data in an organized fashion so that conclusions about the data can be made more readily. One of the ways to organize numerical data is by means of a *frequency table*. As a matter of fact, we have used the frequency table method in most of the experiments in the first three chapters. Here, we shall discuss this method in terms of the frequency table given in Figure 4.1

Frequency Table: It is a table which displays the *frequencies* of all *outcomes* for an experiment.

Frequency of an outcome is the number of times that the outcome occurs. See Prologue, page ii.

Tally marks help to record the outcomes as they occur.

Results of Rolling a Die 30 Times

Outcome	Tally	Frequency (f)
1	̶H̶H̶1̶ 11	7
2	̶H̶H̶1̶	5
3	̶H̶H̶1̶	5
4	̶H̶H̶1̶	4
5	111	3
6	̶H̶H̶1̶ 1	6
		Total = 30

Figure 4.1

The first column displays all six possible outcomes for the experiment. The second column records the occurrence of each outcome by a vertical stroke. The fifth time that a particular outcome occurs we represent it by a diagonal stroke so that the *occurrences of an outcome are organized into groups of five.*

This procedure helps us to keep an accurate count of the frequency of each outcome as recorded in the last column. At the bottom of the column, we have the number 30, which is often referred to as the "Total number of trials," or the sum of all frequencies.

Once we have obtained the frequency for each outcome we may omit the tally column in our presentation of the results for the experiment such as the table in Figure 4.2.

Total number of trials

Results of Rolling a Die 30 Times

Outcome	Frequency
1	7
2	5
3	5
4	4
5	3
6	6
	Total = 30

Figure 4.2

Table with the tally column removed

Since the table in Figure 4.2 shows how the frequencies of the outcomes are distributed, we often refer to it as a *frequency distribution* for the experiment. From the frequency distribution, we can calculate the frequency of an event, which means the number of times that the event occurs. For example, the frequency of the event

Frequency distribution

Frequency of an event

$$B = \{3,4,5\} \text{ is } 5 + 4 + 3 = 12.$$

The frequency of a simple event is often stated as the frequency of an outcome.

Frequency of an outcome

BAR GRAPH

The frequency distribution in Figure 4.2 could also be represented by a bar graph as shown in Figure 4.3.

Bar Graph:
Conclusions about the frequencies of outcomes could be drawn by comparing the lengths of the bars. Note that the horizontal axis is labelled *Outcome*, and the vertical axis *Frequency*.

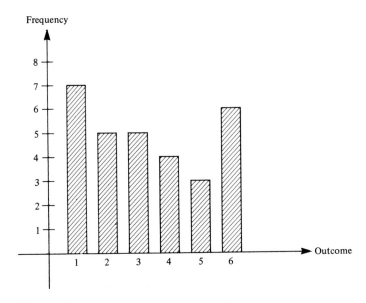

Figure 4.3

How to read the bar graph.

On this graph the outcomes are represented by equally spaced points on the *horizontal axis*, one point for each outcome. The height of the bar for each outcome, indicated by the scale on the *vertical axis*, represents the frequency of the outcome in the experiment. For example, the bar for the outcome 1 has a height of 7 units. This means that the outcome 1 has occurred 7 times in the experiment. It is also the outcome that occurs most frequently, since its corresponding bar is the tallest among the six bars. Which of the six outcomes occurs least frequently?

Frequency: An Approximation for Expected Number.
It is important to realize at this point that the *frequency* of an outcome is not the same as the *expected number* of the outcome. Given a number of trials of a chance experiment, the frequency of an outcome is obtained as a result of *practical observations.* The expected number of an outcome is calculated from the *theoretical probability* of the occurrence of the outcome. In practice, we use *the frequency of an outcome to approximate the expected number* of the outcome. This idea of approximation has been used in the first and third chapters. A comparison of the frequency and the expected number for each of the six outcomes of throwing a die 30 times is given in Figure 4.4.

$E = P(A) \times n$

Difference between Frequency and Expected Number

Difference = Larger − smaller

Outcome	Frequency	Expected Number	Difference between Frequency and Expected Number.
1	7	5	2
2	5	5	0
3	5	5	0
4	4	5	1
5	3	5	2
6	6	5	1

Examples:
1. 7 − 5 = 2
2. 5 − 4 = 1

Figure 4.4

Note that the differences in the last column are *small*, thus showing that *frequency is a good approximation of expected number.*

By permission of John Hart and Field Enterprises Inc.

EXERCISES 4.2

1. Open a telephone book to any page at random, and tally the last digit of the first 100 telephone numbers on the page.

Digit	Tally	Frequency
0		
1		
2		
3		
4		
5		
6		
7		
8		
9		
Total		100

Frequency Distribution of Last Digit from
100 Telephone Numbers

a) Give the frequency of a digit which occurred least frequently.

b) Give the frequency of a digit which occurred most frequently.

c) What is the difference between these two frequencies?

d) Did all the ten digits occur with about equal frequency?

e) Draw the bar graph for this frequency distribution.

Label the horizontal axis *Digit*, and the vertical axis *Frequency*.

For each of the Exercises 2 through 5 below:

a) Calculate the expected number for each outcome to one decimal place.

$E = P(A) \times n$

b) Use the appropriate experimental results which you obtained in the exercises of Chapter 1 to fill in the column "frequency."

2. Flipping a coin 60 times (Experiment I, page 36).

See p. 37

Outcome	Probability	Expected Number	Frequency	The difference between frequency and expected number
Head	$\frac{1}{2}$			
Tail	$\frac{1}{2}$			

3. Throwing a die 60 times (Experiment II, page 38).

See p. 40

Outcome	Probability	Expected Number	Frequency	The difference between frequency and expected number
1	$\frac{1}{6}$			
2	$\frac{1}{6}$			
3	$\frac{1}{6}$			
4	$\frac{1}{6}$			
5	$\frac{1}{6}$			
6	$\frac{1}{6}$			

4. Flipping a penny and a nickel together 30 times (Experiment III, page 47).

See p. 48

Outcome	Probability	Expected Number	Frequency	The difference between frequency and expected number
2 heads	$\frac{1}{4}$			
1 head	$\frac{2}{4}$			
0 heads	$\frac{1}{4}$			

5. Rolling a pair of dice (one green and one red) together 50 times (Experiment IV, page 49).

See p. 50

Outcome	Probability	Expected Number	Frequency	The difference between frequency and expected number
2	$\frac{1}{36}$			
3	$\frac{2}{36}$			
4	$\frac{3}{36}$			
5	$\frac{4}{36}$			
6	$\frac{5}{36}$			
7	$\frac{6}{36}$			
8	$\frac{5}{36}$			
9	$\frac{4}{36}$			
10	$\frac{3}{36}$			
11	$\frac{2}{36}$			
12	$\frac{1}{36}$			

4.3 RELATIVE FREQUENCY: AN APPROXIMATION FOR PROBABILITY

The last column in Figure 4.2 shows that the outcome 1 occurs most frequently (7 times out of 30), and the outcome 5 occurs least frequently (3 times out of 30). On the basis of these results, would you suspect that the die is biased against the outcome 5? How many times would you like to see the outcome 5 occur in order for you to conclude that the die is not biased against the outcome 5?

Rolling a die 30 times

To answer these questions, let us fall back on what we already know about the probability of the outcome 5. We know that if a die is fair, then the outcome 5 has the probability $\frac{1}{6} \approx 17\%$ to occur. This is also the case for every other outcome. Since the total number of tosses in the experiment is 30, and the frequency of the outcome 5 is 3, then the outcome 5 occurs only

$$\frac{3}{30} = \frac{1}{10} = 10\% \text{ of the time.}$$

But the outcome 1 occurs $\frac{7}{30} \approx 23\%$ of the time. Thus, the

$\frac{7}{30} \approx 23\%$

Check this approximation
See Sec. C.2.

outcome 5 occurs about *7% less frequently* than we would expect, and the outcome 1 about *6% more frequently* than expected. The calculated fractions (percentages) for the outcomes 1 and 5 are called their *relative frequencies.*

Definition. *The relative frequency of an outcome is the frequency of the outcome* relative *to the total number of trials.*

Relative Frequency of an outcome

$$\text{Relative Frequency} = \frac{\text{Frequency}}{\text{Total Number of Trials.}}$$

$\text{R.F.} = \frac{f}{n}$

The relative frequency for each of the six outcomes for the die experiment just discussed is shown in Figure 4.5.

Results of Rolling a Die 30 Times

Outcome	Frequency	Relative Frequency
1	7	$\frac{7}{30} \approx 0.23 = 23\%$
2	5	$\frac{5}{30} \approx 0.17 = 17\%$
3	5	$\frac{5}{30} \approx 0.17 = 17\%$
4	4	$\frac{4}{30} \approx 0.13 = 13\%$
5	3	$\frac{3}{30} = 0.10 = 10\%$
6	6	$\frac{6}{30} = 0.20 = 20\%$
Total	30	$\frac{30}{30} = 1.00 = 100\%$

$$\frac{5}{30} = \frac{1}{6} \approx 17\%$$

$$\frac{4}{30} = \frac{2}{15} \approx 13\%$$

See Sec. B.3, B.4, C.2.

Figure 4.5

Rounding error
The difference between a number and its rounded number is called a rounding error.

Note that *the sum of all the relative frequencies must be equal to 1 or 100% (Why?). However, owing to rounding error, the sum may sometimes be slightly less or more than one or 100%.*

We have looked at the differences between the relative frequency and the probability of each of the outcomes 1 and 5. A complete picture of these differences for all the six outcomes is given in Figure 4.6.

Difference between Relative Frequency and Probability

Outcome	Relative Frequency	Probability	Difference between Relative Frequency and Probability
1	23%	17%	6%
2	17%	17%	0%
3	17%	17%	0%
4	13%	17%	4%
5	10%	17%	7%
6	20%	17%	3%

Difference
= Larger − smaller
Examples

1. $23\% - 17\%$
$= \dfrac{23-17}{100}$
$= 6\%$

2. $17\% - 10\%$
$= \dfrac{17-10}{100}$
$= 7\%$

Figure 4.6

The difference between the relative frequency and the probability of the outcome is the greatest for the outcome 5. It is 7%. There is no such difference for the outcomes 2 and 3.

What does Figure 4.6 suggest to us? Can we confidently draw the conclusion that 7% is large enough for us to say that the die is biased against the outcome 5? If so, would you bet against the outcome 5 with your friends on a dollar-for-dollar basis? Probably you might want to play safe by throwing the die another 30 times, and better still by asking someone else to throw the die for you so that you might eliminate your own suspicion that the small frequency of the outcome 5 might be partly due to the particular way you threw the die. Figure 4.7 shows the results of another 30 throws.

7% is the largest of the six differences

Is the die biased?

We roll the die another 30 times.

Results of Rolling a Die 30 Times

Check the conversion from fraction into percent.

See Sec. B.3, C.2.

Outcome	Tally	Frequency	Relative Frequency
1	1111	4	$\frac{4}{30} \approx 0.13 = 13\%$
2	TTTT 1	6	$\frac{6}{30} = 0.20 = 20\%$
3	111	3	$\frac{3}{30} = 0.10 = 10\%$
4	TTTT 11	7	$\frac{7}{30} \approx 0.23 = 23\%$
5	TTTT 1	6	$\frac{6}{30} = 0.20 = 20\%$
6	1111	4	$\frac{4}{30} \approx 0.13 = 13\%$
Total		30	$\frac{30}{30} \approx 0.99 = 99\%$

Figure 4.7

Again, let us compare the relative frequency and the probability for each of the six outcomes.

Difference between Relative Frequency and Probability

Check the differences in the last column.

7% is the largest of the six differences.

Outcome	Relative Frequency	Probability	Difference between Relative Frequency and Probability
1	13%	17%	4%
2	20%	17%	3%
3	10%	17%	7%
4	23%	17%	6%
5	20%	17%	3%
6	13%	17%	4%

Figure 4.8

This time the relative frequency of the outcome 5 is 3% more than its probability (in percent). On the other hand, the relative frequency of the outcome 1 is 4% less than its probability. In view of the first set of results, these results may surprise you a bit. Perhaps you are now not so sure that the die is biased against the outcome 5.

Faced with these 2 sets of results, what can we do so that we can draw some *reliable conclusions* as to the fairness or biasedness of the die? *One way to resolve these conflicting results is to regard them as the results of throwing the die 60 times at one time instead of 30 times twice.* According to this conception, we then have the following frequency distribution for one experiment of rolling a die 60 times.

Draw a conclusion.

Check the differences in the last column. 4% is the largest of the six differences.

Difference between Relative Frequency and Probability

Outcome	Frequency	Relative Frequency	Difference between Relative Frequency and Probability
1	7+4=11	$\frac{11}{60} \approx 18\%$	1%
2	5+6=11	$\frac{11}{60} \approx 18\%$	1%
3	5+3= 8	$\frac{8}{60} \approx 13\%$	4%
4	4+7=11	$\frac{11}{60} \approx 18\%$	1%
5	3+6= 9	$\frac{9}{60} \approx 15\%$	2%
6	6+4=10	$\frac{10}{60} \approx 17\%$	0%

Figure 4.9

Figure 4.9 shows that the largest difference between relative frequency and probability is only about 4%. The first 2 sets of results both have about 7% as the largest difference between relative frequency and probability. The *differences* for *60 throws* is *smaller* in general than those of any of the two sets of 30 throws as seen in the last column of Figures 4.6, 4.8, and 4.9.

In fact, it can be proved in the advanced theory of probability that these differences get smaller and smaller as the number of throws gets larger and larger for a fair die. Therefore, we would expect to see that these differences for 120 throws would be smaller than those for 60 throws. (Test this statement by throwing a fair die 120 times.)

The results of the last column in Figure 4.9 are sufficient evidence for us to say that the die that we have used is a fair one. Of course, we might be wrong, but the probability of our conclusion being incorrect, we believe, is very small on the basis of the evidence collected.

In conclusion, we may say that *the relative frequency of an outcome is a good approximation of its theoretical probability.* The approximation increases its accuracy as the number of trials of the experiment increases.

The notion of the relative frequency of an outcome can be readily extended to relative frequency of more than one outcome or of a compound event.

This fact is called the Law of Large Numbers.

We conclude that the die is a fair one.

$$\frac{f}{n} \longrightarrow P$$

$\frac{f}{n}$ approaches P

Relative Frequency of an event *A*.

Definition. *The relative frequency of an* event *A is the frequency of the event relative to the total number of trials.*

Thus, relative frequency of $A = \dfrac{\text{Frequency of } A}{\text{Total number of trials}}$

For example, the relative frequency for the event $A = \{2, 4, 6\}$ from Figure 4.9 is

Relative Frequency of the event *A*

$$= \frac{11 + 11 + 10}{60}$$

$$= \frac{32}{60}$$

$$= \frac{8}{15}$$

$$\approx .533$$

$$\approx 53\%$$

$$\frac{32}{60} = \frac{8 \times 4}{15 \times 4} = \frac{8}{15}$$

EXERCISES 4.3

1. Using the results that you obtained in Exercise 1 of Exercises 4.2, find the relative frequency of each of the 10 decimal digits.

2. The following table shows the distribution of male and female students across their year of study in a small college of 2,163 students.

	Freshman	Sophomore	Junior	Senior
Male	310	270	190	152
Female	450	310	251	230

Find the relative frequency (in percent) of:

Method for (a)
1. How many male freshmen in the college?
2. The number of male freshmen is what percent of the total student population?

    a)   the male freshmen in the college.
    b)   the female juniors in the college.
    c)   the sophomores in the college.
    d)   the female students in the college.
    e)   the juniors and the seniors in the college.

3. Construct a bar graph for the following frequency distribution:

Year of Study	Number of Students
Freshman	760
Sophomore	580
Junior	441
Senior	382
Total	2163

Label the horizontal axis *Year of Study*, and the vertical axis *Number of Students.*

4. A poll of 116 persons is taken on the question of raising the price of beef. The results of the poll are tabulated below.

	Meat Packers	Housewives	Others
Favor	20	10	5
Oppose	2	50	6
No opinion	1	12	10

Method:
Use relative frequency to approximate the required probability.

If one of the 116 persons is chosen randomly, what is the probability that the chosen person:

    a)   opposes raising the price of beef?
    b)   favors raising the price?
    c)   has no opinion on the question?
    d)   opposes *or* has no opinion?

Name: _____

Experiment VII: *Flipping five coins 50 times*

AIM: *To show that relative frequency is a good approximation of probability.*

Take five coins and flip them together. Count the number of heads that appear and record with a tally opposite this number in the first column of the following table. Repeat this operation for a total of 50 times.

Number of heads	Tally	Frequency	Relative Frequency	Probability	The difference between relative frequency and probability.
0				$\frac{1}{32}$	
1				$\frac{5}{32}$	
2				$\frac{10}{32}$	
3				$\frac{10}{32}$	
4				$\frac{5}{32}$	
5				$\frac{1}{32}$	
Total				$\frac{32}{32} = 1$	

1. Which of the outcomes occurred most frequently?

   _____

2. What is the approximate probability that you will get the outcome that you have given in question 1 when you flip the five coins again?

   _____

3. What is the approximate probability that you will get either one head or two heads?

_____

4. What is the smallest difference between the relative frequency and the true probability of an outcome?

_____

5. What is the largest difference between the relative frequency and the true probability of an outcome?

CONCLUSION: *Have you achieved the aim of this experiment? Explain.*

_____

_____

_____

_____

4.4 THE MEAN OF RELATIVE FREQUENCIES

$$\frac{f_1 + f_2}{2}$$

$$\frac{0.23 + 0.13}{2} = 0.18$$

The relative frequency in Figure 4.9 can also be calculated by dividing the sum of the two relative frequencies of an outcome in Figures 4.5 and 4.7 by two.

For example, the two relative frequencies of the outcome 1 are 0.23 and 0.13. Thus the sum of the two relative frequencies is

$$0.23 + 0.13 = 0.36$$

Dividing 0.36 by 2, we have

$$\frac{0.36}{2} = 0.18$$

The relative frequency 0.18 so obtained is called the *mean* of the two relative frequencies 0.23 and 0.13 of the outcome 1. The other five means are similarly found and are given in Figure 4.10.

Mean

Definition. *The* mean *of a set of numbers is the result found by dividing the sum of the numbers by the number of numbers added.*

Means of Two Relative Frequencies

Outcome	Relative Frequency		Mean of Two Relative Frequencies
1	0.23	0.13	$\frac{0.23+0.13}{2} = 0.18 = 18\%$
2	0.17	0.20	$\frac{0.17+0.20}{2} \approx 0.19 = 19\%$
3	0.17	0.10	$\frac{0.17+0.10}{2} \approx 0.14 = 14\%$
4	0.13	0.23	$\frac{0.13+0.23}{2} = 0.18 = 18\%$
5	0.10	0.20	$\frac{0.10+0.20}{2} = 0.15 = 15\%$
6	0.20	0.13	$\frac{0.20+0.13}{2} \approx 0.17 = 17\%$

Figure 4.10

The slight discrepancies between the mean of the two relative frequencies in Figure 4.10 and the relative frequency in Figure 4.9 for the two outcomes 2 and 3 are the results of the cumulative effects of the rounding procedure that we used in finding the two relative frequencies and their mean for each outcome.

Averages: Mean, Median, and Mode. You may have noted that the idea of the mean that we have just introduced is very similar to the idea commonly known as "average." In common usage, the word *"average" has a number of meanings.* For example, it has different meanings in the following three statements:

1. The *average* shoe size of a man in the United States is *nine.*

2. The *average* size sweater is easier to find in a store.

3. The *average* expenses of a family of three are *$600* a month.

In the first statement, the word "average" probably means that even though the shoe size may vary from man to man, *the largest number of men* have size nine. In this usage, the word "mode" is preferable to "average" in statistics.

Average

Mode

Definition. A mode *of a set of numbers is a number that occurs* most *frequently.*

In the second statement, the word "average" probably means "median," since sweaters are usually classified into three sizes: small, medium, and large. The word "average" in this usage has the meaning of being the *middle.*

Median

Definition. *The* median *of a set of numbers is the middle number when the numbers are arranged in order of size.*

In the third statement, the meaning of the word "average" is really that of the mean, since family monthly expenses are determined by dividing the total monthly expenses by the number of months taken.

Let us consider a few examples to further clarify the meanings of the three words, mean, median, and mode, which are commonly replaced by the word "average."

Examples:

1. Find the mean, median, and mode of the set of numbers: 3,7,9,11,6,5,12,3,8.

Solution:

a) Since there are 9 numbers in the set of given numbers, we, therefore, have

$$mean = \frac{3 + 7 + 9 + 11 + 6 + 5 + 12 + 3 + 8}{9}$$

$$= \frac{64}{9}$$

$$= 7\frac{1}{9}$$

See Sec. A.8, A.9.

Mean $= 7\frac{1}{9}$

b) First arrange the numbers from the smallest to the largest (or from the largest to the smallest). Thus, we have 3,3,5,6,7,8,9,11,12.

$3 < 5 < 6 < 7 < \ldots < 12$

Since the number 7 is in the middle position when the numbers are arranged in order of size, thus, the *median* is 7.

Median $= 7$

c) Since the number 3 occurs most frequently among the given numbers, then the *mode* is ɔ.

Mode = 3

2. Find the mean, median, and mode of the set of numbers: 1.02, 3.84, 2.0, 5.1, 3.0, 4.3.

Solution:

a) Since there are six numbers in the set of given numbers, we then have

$$Mean = \frac{1.02 + 3.84 + 2.0 + 5.1 + 3.0 + 4.3}{6}$$

See Sec. B.5, B.7

$$= \frac{19.26}{6}$$

$$= 3.21$$

Mean = 3.21

b) Rearranging the numbers from the smallest to the largest, we have

1.02, 2.0, 3.0, 3.84, 4.3, 5.1.

$1.02 < 2.0 < 3.0 < \ldots < 5.1$

Since there are two numbers in the middle positions, we would define the median to be the mean of the two middle numbers. That is:

$$Median = \frac{3.0 + 3.84}{2}$$

$$= \frac{6.84}{2}$$

$$= 3.42$$

Median = 3.42

c) Since all the given numbers occur with the *same frequency,* we would prefer to say that none of them is a mode.

No mode

Note that the mean and the median of a set of numbers always exist, but this is not necessarily so for the mode.

A set of numbers may not have a mode

Question:

Can you give a set of numbers which has more than one mode?

1,1,2,2,3,4,5

EXERCISES 4.4

1. An experiment of flipping a coin 20 times is repeated 5 times. The results are given in the table below:

Number of Flips	Relative Frequency	
	(Head)	(Tail)
First 20	.45	.55
Second 20	.50	.50
Third 20	.40	.60
Fourth 20	.45	.55
Fifth 20	.60	.40

a) What is the best approximation for the probability of getting a head?

b) What is the best approximation for the probability of getting a tail?

c) Judging from the two approximated probabilities, can one conclude that the coin used in the experiment is a *fair* one?

2. Find the mean and the median for each of the following sets of numbers:

a) 100, 140, 139, 157, 98, 175, 149, 130, 135.

b) 1.24, 3.12, 0.19, 5.1, .7, 4.17, 5.23.

c) $\frac{1}{2}, \frac{2}{3}, \frac{3}{4}, \frac{4}{5}, \frac{5}{6}$.

d) $2\frac{1}{3}, 4\frac{3}{5}, 5, 3\frac{2}{5}, 3, \frac{2}{3}$.

e) 9%, 12%, 20%, 30%, 50%, 36%.

Comparison of decimals

Comparison of fractions

See Sec. A.3

3. A class of 12 students weighed a piece of metal on a chemical balance with their results given below:

.1531, .1533, .1531, .1535, .1529, .1534,
.1534, .1532, .1534, .1535, .1528, .1529.

a) What is the mean weight of the metal?

b) What is the median weight of the metal?

c) What is the modal weight of the metal?

d) Is the mean weight a good estimate of the weight of the metal? Explain your answer.

4. The table below shows the fuel consumption of a car for 3 days.

Gasoline (in gallons)	Distance traveled (in miles)
2.1	48.3
3.2	72.7
4.5	90.8

What is the mean mileage of this car per gallon of gasoline?

5. Count off the first 50 words in the introduction to this chapter (excluding mathematical symbols). Tally the length of these 50 words, the length of a word being defined as the number of letters of the word.

Length of words (L)	Tally	Frequency (f)	$L \, X \, f$
1			
2			
3			
4			
5			
6			
7			
8			
9			
10			
11			
12			
13			
14			
Total			

Frequency Distribution of Lengths of 50 words.

a) What is the sum of the lengths of these 50 words?
b) What is the mean length of a word among these 50 words?
c) Draw the bar graph for this frequency distribution.

Division of decimals:

$$\frac{2.25}{1.5} = \frac{2.25 \times 10}{1.5 \times 10}$$
$$= \frac{22.5}{15}$$
$$= 1.5$$

See Sec. B.7

Label the horizontal axis *Length of words*, and the vertical axis *Frequency*.

4.5 UNEQUALLY LIKELY OUTCOMES

Up to this point, we have used relative frequency to determine whether or not we could assume all outcomes of an experiment to be equally likely to occur. We have seen that, in the absence of any effort to influence the outcome, the relative frequency is a good approximation of the probability of an outcome.

We shall now apply the relative frequency method to experiments or situations where the equal likelihood of outcomes is questionable, or the most feasible way to find a probability is to use relative frequency to estimate it.

Let us examine a simple experiment whose outcomes, we believe, are *not* equally likely to occur.

Thumbtack experiment

Take an ordinary thumbtack and throw it into the air. Let it drop onto a table, where it bounces before coming to rest. When it finally comes to rest, it either lands point up  or point down . Let us use the small letter u to denote the outcome point-up, and the small letter d for the outcome point-down. Thus, a sample space for this experiment is

$$S = \{u, d\}$$

Are the outcomes u and d equally likely to occur?

Is $S = \{u, d\}$ the primitive sample space? In other words, is $P(\{u\}) = P(\{d\})$?

Our experience with thumbtacks may lead us to believe that one outcome is more likely to occur than the other. Some may say $P(\{u\}) > P(\{d\})$, and others $P(\{d\}) > P(\{u\})$. Still other people may say that it depends on the type of thumbtack that we have.

The key question.

The question is how do we find out whether $P(\{u\}) > P(\{d\})$ or the reverse is true for a particular thumbtack?

Clearly, if we know the numerical value of $P(\{u\})$ and $P(\{d\})$, then the question is answered. Before we proceed to find their probabilities, let us note at this point that the two simple events $\{u\}$ and $\{d\}$ are *complementary* to each other. So it suffices to find the probability of one of them, since the probability of the other can be calculated from the formula:

$\{u\}$ and $\{d\}$ are complementary

$$P(\{u\}) = 1 - P(\{d\}) \text{ or}$$
$$P(\{d\}) = 1 - P(\{u\}).$$

$P(\bar{A}) = 1 - P(A)$

How can we find the numerical value of $P(\{u\})$?

It is hard, if not impossible, to know the true value of $P(\{u\})$. However, it is easy to obtain the *approximate value* of $P(\{u\})$ by the *relative frequency method* that we have already discussed.

Take the thumbtack and throw it into the air a large number of times, say 50 times. The frequency of each outcome is given in the frequency table below.

Results of Throwing a Thumbtack 50 Times

Outcome	Tally	Frequency	Relative Frequency
u	THL THL THL THL THL THL 1	31	$\dfrac{31}{50} = .62$
d	THL THL THL 1111	19	$\dfrac{19}{50} = .38$

Figure 4.11

Judging from the two relative frequencies, .62 for the outcome u and .38 for the outcome d, we would conclude that

$$P(\{u\}) > P(\{d\}). \text{ Why?}$$

Of course, we are aware that we might be wrong in our conclusion. To be more certain of our conclusion, we may repeat the same experiment another 50 times, and see if the new results are consistent with the previous results. Or, we may combine the two sets of experimental results into one bigger set, and then calculate the relative frequencies of the two outcomes in order to draw a more reliable conclusion about them.

Once we accept the empirical probabilities $P(\{u\}) = .62$ and $P(\{d\}) = .38$, then we have to accept the consequence that the sample space $S = \{u,d\}$ is *not* a primitive sample space. This is not a serious matter, because we already know the probabilities of all outcomes in S.

Let us now consider a situation where the relative frequency method seems to be indispensable.

By actually tossing the thumbtack, we approximate the value of $P(\{u\})$.

$$\begin{array}{r} .62 \\ 50\,\overline{)31.0} \\ 30\ 0 \\ \hline 1\ 00 \\ 1\ 00 \\ \hline\hline \end{array}$$

See Sec. B.3

$$P(\{u\}) > P(\{d\})$$

Empirical probability

$S = \{u,d\}$ not a primitive sample space

There are many situations where the relative frequency method must be used to determine the probability.

Success of a new drug.

Suppose we are told that 230 out of 250 heart patients have immediate improvement upon the injection of a new drug. Therefore, the relative frequency of improvement in heart patients is

$$\frac{230}{250} = \frac{23}{25} = .92 \text{ or } 92\%$$

On the basis of this evidence, we may conclude that the probability that the new drug will improve the condition of a heart patient is 92%.

$P(\overline{A}) = 1 - P(A)$

Question: What is the probability that the new drug will *not* improve the condition of a heart patient?

Examples:

1. A railway worker in a small town has observed that in the past year (365 days) a cargo train which comes into the town once every day has been late 30 times. Approximate the probability that the train will arrive on time.

Solution:

$$\begin{array}{r} .917 \approx .92 \\ 365 \overline{)335.00} \\ 328\,5 \\ \hline 6\,50 \\ 3\,65 \\ \hline 2\,850 \\ 2\,555 \\ \hline 295 \end{array}$$

See Sec. B.3, B.4.

The relative frequency for the train to arrive late in a year $= \frac{30}{365}$, and thus the relative frequency for the train to arrive on time in a year is

$$\frac{365 - 30}{365} = \frac{335}{365} \approx 0.92 = 92\%.$$

Hence the approximate probability that the train will arrive on time is 92%, which means that the train has a high degree of punctuality.

Life insurance

2. One of the working assumptions of life insurance companies is that what will happen in the near future will be similar to what has happened in the recent past. Thus, they collect and use sets of observations, such as the table of mortality on

page 156, to aid in their prediction of the probability of the death of people of a given age in a particular year. The probability, as we have discussed, is approximated by relative frequency.

See Sec. C.2.

According to the Mortality Table, at age 20 the probability that death will occur within a year, is approximately

$$\frac{1.79}{1000} = 0.00179 \approx 0.2\%.$$

$0.00179 \approx 0.002$
$\approx .2\%$

This is much less than 1%.

However, at age 98 the probability that death will occur within a year is approximately equal to

$$\frac{668.15}{1000} = 0.66815 \approx 67\%.$$

$0.66815 \approx 0.67 = 67\%$

At age 99 such a probability is almost 1. That is, a person 99 years old is almost certain to die within a year.

Suppose we want to know the probability that people in a certain age group will die within a year. We could use the *mean of the relative frequencies* of the ages included in the given age group to approximate it. For example, consider the 20-25 age group; the people in this age group will die within a year with an approximate probability

Mean of relative frequencies.

$$\frac{0.00179 + 0.00183 + 0.00186 + 0.00189 + 0.00191 + 0.00193}{6}$$

$$= \frac{0.01121}{6} = 0.00187 \approx 0.2\%$$

Check the addition and division.

See Sec. B.5, B.7.

Consider another age group, 70-75. The probability that the people in this age group will die within a year is approximately equal to

$$\frac{0.04979 + 0.05415 + 0.05865 + 0.06326 + 0.06812 + 0.07337}{6}$$

$$= \frac{0.36734}{6} \approx 0.06122 \approx 6.1\%$$

Check the addition and division.

$0.06122 \approx 0.061 = 6.1\%$

This is considerably higher than the probability for the 20-25 age group, which is only 0.2%.

Mortality Table (Commissioners 1958 Standard Ordinary)

Age	Number Living	Deaths Each Year	Deaths Per 1,000	Age	Number Living	Deaths Each Year	Deaths Per 1,000	Age	Number Living	Deaths Each Year	Deaths Per 1,000
0	10,000,000	70,800	7.08	34	9,396,358	22,551	2.40	67	6,355,865	241,777	38.04
1	9,929,200	17,475	1.76					68	6,114,088	251,835	41.68
2	9,911,725	15,066	1.52	35	9,373,807	23,528	2.51	69	5,859,253	267,241	45.61
3	9,896,659	14,449	1.46	36	9,350,279	24,685	2.64				
4	9,882,210	13,835	1.40	37	9,325,594	26,112	2.80	70	5,592,012	278,426	49.79
				38	9,299,482	27,991	3.01	71	5,313,586	287,731	54.15
5	9,868,375	13,322	1.35	39	9,271,491	30,132	3.25	72	5,025,855	294,766	58.65
6	9,855,053	12,812	1.30					73	4,731,089	299,289	63.26
7	9,842,241	12,401	1.26	40	9,241,359	32,622	3.53	74	4,431,800	301,894	68.12
8	9,829,840	12,091	1.23	41	9,208,737	35,362	3.84				
9	9,817,749	11,879	1.21	42	9,173,375	38,253	4.17	75	4,129,906	303,011	73.37
				43	9,135,122	41,382	4.53	76	3,826,895	303,014	79.18
10	9,805,870	11,865	1.21	44	9,093,740	44,741	4.92	77	3,523,881	301,997	85.70
11	9,791,005	12,047	1.23					78	3,221,884	299,829	93.06
12	9,781,958	12,325	1.26	45	9,048,999	48,412	5.35	79	2,922,055	295,683	101.19
13	9,769,633	12,896	1.32	46	9,000,587	52,473	5.83				
14	9,756,737	13,562	1.39	47	8,948,114	56,910	6.36	80	2,626,372	288,848	109.98
				48	8,891,204	61,794	6.95	81	2,337,524	278,983	119.35
15	9,743,175	14,225	1.46	49	8,829,410	67,104	7.60	82	2,058,541	265,902	129.17
16	9,728,950	14,983	1.54					83	1,792,639	249,858	139.38
17	9,713,967	15,737	1.62	50	8,762,306	72,902	8.32	84	1,542,781	231,433	150.01
18	9,698,230	16,390	1.69	51	8,689,404	79,160	9.11				
19	9,681,840	16,846	1.74	52	8,610,244	85,758	9.96	85	1,311,348	211,311	161.14
				53	8,524,486	92,832	10.89	86	1,100,037	190,108	172.82
20	9,664,994	17,300	1.79	54	8,431,654	100,337	11.90	87	909,929	168,455	185.13
21	9,647,694	17,655	1.83					88	741,474	146,997	198.25
22	9,630,039	17,912	1.86	55	8,331,317	108,307	13.00	89	594,477	126,303	212.46
23	9,612,127	18,167	1.89	56	8,223,010	116,849	14.21				
24	9,593,960	18,324	1.91	57	8,106,161	125,970	15.54	90	468,174	106,809	228.14
				58	7,980,191	135,663	17.00	91	361,365	88,813	245.77
25	9,575,636	18,481	1.93	59	7,844,528	145,830	18.59	92	272,552	72,480	265.93
26	9,557,155	18,732	1.96					93	200,072	57,881	289.30
27	9,538,423	18,981	1.99	60	7,698,698	156,592	20.34	94	142,191	45,026	316.66
28	9,519,442	19,324	2.03	61	7,542,106	167,736	22.24				
29	9,500,118	19,760	2.08	62	7,374,370	179,271	24.31	95	97,165	34,128	351.24
				63	7,195,099	191,174	26.57	96	63,037	25,250	400.56
30	9,480,358	20,193	2.13	64	7,003,925	203,394	29.04	97	37,787	18,456	488.42
31	9,460,165	20,718	2.19					98	19,331	12,916	668.15
32	9,439,447	21,239	2.25	65	6,800,531	215,917	31.75	99	6,415	6,415	1000.00
33	9,418,208	21,850	2.32	66	6,584,614	228,749	34.74				

Reading the table:

1. Age 0 means the age range from birth until the first birthday. For this age range, deaths per 1000 is $7.08 = \dfrac{7.08}{1000}$.

2. Age 1 means the age range from the first birthday until the second. For this age range, deaths per 1000 is $1.76 = \dfrac{1.76}{1000}$.

EXERCISES 4.5

1. Describe a statistical procedure that you would use to determine the probabilities of the six outcomes of a loaded die?

2. Of 1,000 persons now living at age 20, statistics show that 623 will be alive at age 60. What is the probability a person aged 20 will live to be 60? That he will die before he is 60?

3. With reference to the mortality table given on page 156, find the approximate probability that:

   a) a person 20 years old will live to be 21.

   b) a person 70 years old will live to be 71.

   c) a person in the 25-30 age group will be alive for a year.

   d) a person in the 40-45 age group will be alive for a year.

Use the relative frequency to approximate the probabilities.

$$P(\overline{A}) = 1 - P(A)$$

4. In Exercise 3 of Exercises 4.3, if a student is randomly selected from the group of 2,163 students, what is the probability that the student selected is:

   a) female?

   b) a senior?

   c) a female senior?

   d) female, given that she is a junior?

   e) a junior, given that she is a female?

   f) a freshman or sophomore?

   g) a female freshman or a male sophomore?

   h) male or a junior?

$n(S) = 2163$

$A = \{\text{ female students }\}$
$B = \{\text{ seniors }\}$
$C = \{\text{ female seniors }\}$
What is the sample space for (d)?
What is the sample space for (e)?

5. You are challenged to guess the number of red balls in a bag. Suppose that you are allowed to put 20 green balls into the bag and mix thoroughly, and then draw eight balls at a time from the bag. If you discover that there are only two green balls and the other six are red, then what is a good guess of the number of red balls in the bag?

Relative frequency method

SUMMARY

Approximation Techniques:

1. Frequency of an event is used to approximate the expected number of the event.

2. Relative frequency of an event is used to approximate the probability of the event. The approximation increases in accuracy as the number of trials of an experiment increases.

Terms and Symbols:

Terms	Symbols	Page
Frequency Distribution		130
Bar Graph		132
Theoretical frequency (Expected number)	E	133
Relative Frequency of an outcome.	$\dfrac{f}{n}$	137
Relative frequency of an event.	$\dfrac{f}{n}$	142
Average		147
Mean		146
Median		148
Mode		148

REVIEW EXERCISES

1. Four coins are tossed 30 times; the number of heads obtained each time is recorded as follows:

$$3, 2, 3, 1, 1, 2,$$
$$0, 1, 1, 2, 4, 2,$$
$$1, 2, 2, 1, 2, 3,$$
$$3, 2, 1, 2, 1, 1,$$
$$2, 1, 2, 3, 0, 2.$$

Note that on the first toss 3 heads and 1 tail were obtained; on the second toss 2 heads and 2 tails, and so on.

a) Make a frequency distribution of the number of heads. Label the three columns: "Number of heads," "Tally," and "Frequency."

b) Construct a bar graph for the frequency distribution.

c) Use the frequency distribution, and find the mode (or the number of heads that occurred most frequently in the 30 tosses).

d) Is the mode given in (c) expected? Why?

e) Find the difference between the frequency and the expected number of each outcome.

f) Find the difference between the relative frequency and the probability of each outcome.

g) What is the smallest difference in f?

h) What is the largest difference in f?

i) What is your conclusion about your observations and the mathematical results of tossing 4 coins 30 times?

2. Consider the bar graph for Cadillac sales for each month during 1973.

   a) Which month had the least sales? This represents what percent of the total sales for the year 1973?

   b) Which month had the most sales? This represents what percent of the total sales for the year 1973?

   c) Had the sales in 1973 gone up or down as compared to 1972? By what percent?

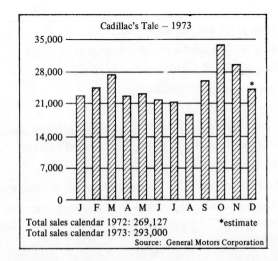

3. Use the Infant Mortality Table to answer the following questions.

Infant Mortality Table

Year	Births	Deaths
1960	153,396	4,012
1970	139,226	3,003

a) What was the approximate probability of an infant born in 1960 surviving the first year?

b) What was the approximate probability of an infant born in 1970 surviving the first year?

c) Has the probability of an infant surviving the first year been increased or decreased from 1960 to 1970? By how much?

d) In a certain hospital, there were 2100 infants born in 1970, how many of them would you expect to survive the first year?

4. Use the Mortality Table to answer the following questions.

Mortality Table

Age	Annual deaths per 100
5	1.35
15	1.46
25	1.93
35	2.51
45	5.35
55	13.00

a) Draw a bar graph for this mortality table.

b) How many annual deaths would you expect in:

(i) a population of 1,000,000 5-year-olds?

(ii) a population of 100,000 25-year-olds?

(iii) a population of 10,000 55-year-olds?

5. A survey of 15 men working in a firm found that 10 earned an annual salary of $8000 each, 3 earned $9000, and 2 earned $100,000 each.

a) Find the mean, median, and modal salary.

b) Why does the mean differ so much from the median and the mode?

c) In reporting the "average" salary of a working man, the mean salary was chosen to be the average salary. Is the choice reasonable? Give your reasons.

6. The following table shows the frequency distribution of 70 scores on a math test.

Score	Frequency
10	2
11	4
12	6
13	7
14	9
15	10
16	13
17	8
18	6
19	3
20	2

a) What is the mean score of the test?

b) Using the table, find the median and the modal scores:

c) Which of the three scores, mean, median, or mode, do you think should be used as a passing score for the test? Give your reasons.

7. Two teams A and B have played each other 14 times. Team A won 9 games, and team B 5 games. They will play again next week. Bob offers to bet $6.00 on team A while you bet $4.00 on team B. The winner gets the $10.00. Is the bet fair to you in view of the past records of the two teams? Explain your answer.

8. A wheel is divided into eight unequal sections numbered 1 through 8. The spinner is spun, and the arrow finally comes to a random stop on one of the sections (if it stops on a line, the spinner is spun again).

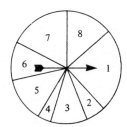

a) What is a sample space for this experiment?

b) At which of the sections is the arrow *most* likely to stop?

c) At which of the sections is the arrow *least* likely to stop?

d) Do the outcomes of this experiment have the same probability of occurring?

e) Describe a procedure by which you can approximate the probability that the arrow stops at section 4.

9. Four groups of 10 college students each are asked to throw a dart at a target, which is placed at a distance of 20 feet away. The target has four circular regions denoted by the numerals 1, 2, 3, and 4. Assume that a throw which misses the target or strikes the circle dividing the regions is not counted. The results of the 40 throws are as follows:

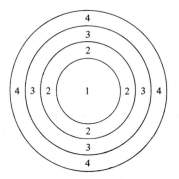

Region	Group A	Group B	Group C	Group D
1	1	0	2	0
2	2	3	1	2
3	4	5	4	6
4	3	2	3	2

a) What is a sample space for the experiment of throwing a dart once?

b) Is it accurate to say that a college student randomly chosen has the same probability (chance) of striking any of the four regions?

c) From the above results of the 40 throws, what is the probability that a randomly chosen college student strikes region 1? Region 2?

d) What is the probability that he will not strike regions 1 or 2?

e) What should he pay for a throw if the prize for striking region 1 is $2.00, that for region 2 is $1.00, and nothing for regions 3 and 4?

Appendices

Appendix A
Fractions

In order to understand the work that we will be doing in probability, it will be necessary to have a good grasp of what fractions are and how to perform operations on them (for example, how to *add* fractions). These concepts will be reviewed in this appendix so that you can refresh your memory if you have not used fractions for some time.

A.1. WHAT IS A FRACTION?

The word "fraction" is a derivative of the Latin word *fractio* meaning "to break." And this is the way the word is most commonly used. For example, if someone were to divide (or "break") a pie evenly among four people then each person would get *a fraction* of the pie. In this case, each person would get *one* piece of the *four* pieces. Therefore, we say that each person would get *one fourth* of the pie and we write this fraction as $\frac{1}{4}$. (See the diagram.)

Each person gets $\frac{1}{4}$.

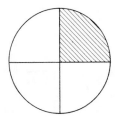

Shaded area:

$\frac{1}{4}$ of the pie

Light area:

$\frac{3}{4}$ of the pie

The number on the "bottom" of the fraction indicates that the pie was divided into *four* parts. The number on the "top" indicates that the person got just *one* of the four parts. For example, if you divided the pie into *five* equal parts and gave Jane *two* of the parts, then the fraction of the pie that she got would be $\frac{2}{5}$, *two fifths*. (See the diagram.)

Jane gets $\frac{2}{5}$

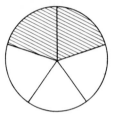

Shaded area:
$\frac{2}{5}$ of the pie

Light area:
$\frac{3}{5}$ of the pie

numerator
denominator

The number on the "top" of a fraction is called the *numerator* and the number on the "bottom" the *denominator*. So, for the fraction $\frac{2}{5}$, 2 is the numerator and 5 is the denominator.

EXERCISES A.1

1. In each of the following diagrams, the shaded part represents *what* fraction of the whole area?

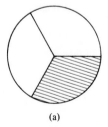

(a)

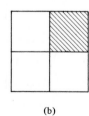

(b)

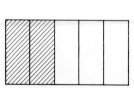

(c)

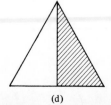

(d)

2. In each of the following two diagrams shade the *fraction* of the area indicated.

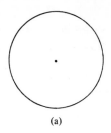

(a)

(b)

a) Shade $\frac{3}{8}$ of the area.

b) Shade $\frac{1}{4}$ of the area in *two* different ways.

A.2 DIFFERENT NAMES FOR THE SAME FRACTION (EQUIVALENT FRACTIONS)

We know that there are many different ways to represent a number. For example, the number *six* can be represented in each of the following ways:

$$6, 4 + 2, 9 - 3, 3 \times 2, 12 \div 2, \frac{6}{1}, \text{ etc.}$$

Of course, in order to avoid confusion we *usually* use the numeral 6 to represent the number six.

numeral

Since a fraction is a number, it can also be represented in various ways. For example, the fraction $\frac{2}{3}$ can be represented by the following

$$\frac{4}{6}, \frac{6}{9}, \frac{12}{18}, \frac{40}{60}, \text{ etc.}$$

$$\frac{2}{3} = \frac{4}{6} = \frac{12}{18} = \text{ etc.}$$

Each of these numerals represents the number $\frac{2}{3}$. This is so because of the following principle:

If the numerator and denominator of a fraction are multiplied by the same number, the value of the fraction remains the same.

Let us look more closely at this principle. Using the fraction $\frac{2}{3}$, multiply the numerator and denominator by 2.

$$\text{That is, } \frac{2 \times 2}{3 \times 2} = \frac{4}{6}$$

$$\text{Therefore, } \frac{4}{6} = \frac{2}{3}.$$

Why is this so?

$\frac{2}{2} = 1$

Because we have multiplied by $\frac{2}{2}$ which is the same as 1, and we know that if we multiply a number by 1 the product is the same as the original number. For example,

$$6 \times 1 = 6.$$

Equivalent Fractions

Numerals such as $\frac{2}{3}$ and $\frac{4}{6}$ which represent the same fraction are called Equivalent Fractions.

EXAMPLE 1

$\frac{2}{7} = ? = ?$

Write two fractions that are equivalent to $\frac{2}{7}$.

Solution:

$$\frac{2 \times 3}{7 \times 3} = \frac{6}{21} \text{ (multiply numerator and denominator by 3)}.$$

$$\frac{2 \times 5}{7 \times 5} = \frac{10}{35} \text{ (multiply numerator and denominator by 5)}.$$

$\frac{2}{7} = \frac{6}{21} = \frac{10}{35}$

$\frac{6}{21}$ and $\frac{10}{35}$ are equivalent to $\frac{2}{7}$.

EXAMPLE 2

Write a fraction with a denominator of 20 that is equivalent to $\frac{3}{4}$.

$$\frac{3}{4} = \frac{?}{20}$$

Solution:

We know that we must multiply the numerator and denominator of $\frac{3}{4}$ by the *same number* to get an equivalent fraction. 4 must be multiplied by what number to get 20? The answer is $20 \div 4 = 5$. (We must multiply by 5.)

$$4 \times 5 = 20$$

$$\text{So, } \frac{3 \times 5}{4 \times 5} = \frac{15}{20}$$

Therefore, $\frac{15}{20}$ is the fraction with denominator of 20 that is equivalent to $\frac{3}{4}$.

$$\frac{3}{4} = \frac{15}{20}$$

EXERCISES A.2

Write *two* other fractions that are equivalent to each of the following fractions:

1. $\frac{1}{3}$

2. $\frac{2}{5}$

3. $\frac{3}{8}$

4. $\frac{5}{11}$

In each of the following, find the missing numerator or denominator so that the new fraction is equivalent to the given fraction.

5. $\frac{1}{7} = \frac{?}{21}$

8. $\frac{5}{8} = \frac{?}{48}$

6. $\frac{2}{3} = \frac{?}{6}$

9. $\frac{5}{6} = \frac{20}{?}$

7. $\frac{3}{5} = \frac{9}{?}$

10. $\frac{7}{25} = \frac{?}{100}$

11. Write a fraction with a denominator of 42 that is equivalent to $\frac{4}{7}$.

12. Change $\frac{2}{9}$ to an equivalent fraction with denominator 36.

A.3 COMPARING FRACTIONS

How do we compare fractions? That is, if we have two fractions such as $\frac{3}{5}$ and $\frac{4}{7}$ – how do we decide which is the larger and which is the smaller number?

The key point to remember is that in order to *compare* fractions the fractions must have the same *base*. That is,

The denominators must be the same.

This principle will also apply when we add and subtract fractions because things cannot be combined unless there is a *basis for comparison*. For fractions this base is a common denominator.

For comparing numbers, two symbols are used (besides the =)

$<$ less than

$>$ greater than

$<$ is read *"less than."*

Thus, $2 < 5$ is read "Two is less than five."

$>$ is read *"greater than."*

Thus, $6 > 4$ is read "Six is greater than four."

EXAMPLE 1

Which is the larger number:

$$\frac{3}{5} \text{ or } \frac{4}{7}\text{?}$$

Solution:

To change $\frac{3}{5}$ and $\frac{4}{7}$ to equivalent fractions with the same denominator, what denominator will be used? Multiply the two denominators and use the *product* as the new denominator.

$$5 \times 7 = 35.$$

$\frac{3}{5} = \frac{?}{35}$

$\frac{4}{7} = \frac{?}{35}$

So, $\frac{3 \times 7}{5 \times 7} = \frac{21}{35}$ (multiply by $\frac{7}{7}$)

$\frac{4 \times 5}{7 \times 5} = \frac{20}{35}$ (multiply by $\frac{5}{5}$)

That is, $\frac{3}{5} = \frac{21}{35}$ and $\frac{4}{7} = \frac{20}{35}$.

Now, which is larger: $\frac{20}{35}$ or $\frac{21}{35}$?

When the denominators are the same we simply compare the numerators.

$$\text{So, } \frac{21}{35} > \frac{20}{35}$$

$$\text{Therefore, } \frac{3}{5} > \frac{4}{7} \qquad\qquad \frac{3}{5} > \frac{4}{7}$$

EXAMPLE 2

Which number is smaller:

$$\frac{4}{11} \text{ or } \frac{7}{20}?$$

Solution:

1. Multiply the denominators.

$$11 \times 20 = 220$$

2. Change each fraction to an equivalent fraction with denominator of 220.

$$\frac{4 \times 20}{11 \times 20} = \frac{80}{220} \qquad\qquad \frac{4}{11} = \frac{?}{220}$$

$$\frac{7 \times 11}{20 \times 11} = \frac{77}{220} \qquad\qquad \frac{7}{20} = \frac{?}{220}$$

We know $77 < 80$, so $\frac{77}{220} < \frac{80}{220}$

$$\text{and } \frac{7}{20} < \frac{4}{11} \qquad\qquad \frac{7}{20} < \frac{4}{11}$$

Therefore, $\frac{7}{20}$ is the smaller fraction.

EXERCISES A.3

For each pair of fractions, determine which represents the larger number. Use the symbols "less than" (<), "greater than" (>), or "equals" (=) in your answer.

1. $\frac{1}{2}$ and $\frac{3}{4}$

2. $\frac{5}{8}$ and $\frac{5}{7}$

3. $\frac{5}{9}$ and $\frac{7}{9}$

4. $\frac{3}{5}$ and $\frac{2}{3}$

5. $\frac{2}{3}$ and $\frac{6}{9}$

6. $\frac{1}{2}$ and $\frac{8}{15}$

7. $\frac{7}{8}$ and $\frac{9}{10}$

8. $\frac{10}{13}$ and $\frac{15}{17}$

List the fractions in each of the following groups in *ascending* order. (That is, write the smallest fraction first, then the next larger one, and so on until the largest one is written last.)

9. $\frac{1}{2}, \frac{3}{4}, \frac{5}{7}$

10. $\frac{3}{8}, \frac{1}{3}, \frac{2}{5}$

11. $\frac{2}{3}, \frac{3}{5}, \frac{2}{9}, \frac{4}{7}$

12. Two stores, Joe's Appliances and Al's Discount House, both sold a particular television set for the same price. Now, each store is having a sale. Joe's is selling the set at "$\frac{1}{3}$ off." Al's is selling it at "$\frac{1}{4}$ off." Where would you buy the set and why?

A.4 REDUCING A FRACTION TO SIMPLEST FORM (LOWEST TERMS)

We have seen in Section A.2 that if we take a fraction such as $\frac{3}{5}$ and multiply numerator and denominator by the same number, we get an equivalent fraction. For example,

$$\frac{3 \times 6}{5 \times 6} = \frac{18}{30}.$$

$$\frac{3}{5} = \frac{18}{30}$$

So, $\frac{3}{5}$ and $\frac{18}{30}$ are two numerals for the same fraction.

If we were to *reverse* this process, we would also get an equivalent fraction. That is, if we *divide* the numerator and denominator by the same number (except 0), we get an equivalent fraction. For example, if we start with the fraction $\frac{10}{15}$ we know that $\frac{20}{30}$ is an equivalent fraction because

$$\frac{10 \times 2}{15 \times 2} = \frac{20}{30}.$$

$$\frac{10}{15} = \frac{20}{30}$$

But we also know that 5 divides 10 and 15 evenly;

$$\frac{10 \div 5}{15 \div 5} = \frac{2}{3}.$$

$$\frac{10}{15} = \frac{2}{3}$$

So, $\frac{2}{3}$ is equivalent to $\frac{10}{15}$.

A number that divides another number evenly is called a *factor* of the second number. So, in this example, 5 is a factor of 10 and a factor of 15.

Factor

When we change a fraction to an equivalent fraction by dividing numerator and denominator by a common factor, we say that we have reduced *the original fraction.*

Reducing a fraction.

In the previous example, $\frac{10}{15}$ was reduced to $\frac{2}{3}$.

Can the fraction $\frac{2}{3}$ be reduced? That is, is there any common factor (besides 1) for 2 and 3? The answer, of course, is no, and so we say that the fraction $\frac{2}{3}$ is in its *simplest form*. (It can also be said that the fraction $\frac{2}{3}$ is in its *lowest terms*.)

Simplest form or lowest terms

EXAMPLE

$\frac{20}{32} = ?$

Reduce the fraction $\frac{20}{32}$ to its simplest form.

Solution:

Both 20 and 32 can be divided evenly by 2. (2 is a common factor.)

$$\text{So, } \frac{20 \div 2}{32 \div 2} = \frac{10}{16}.$$

2 divides 10 and 16 evenly.

But we notice now that both 10 and 16 also have 2 as a common factor.

$$\text{So, } \frac{10 \div 2}{16 \div 2} = \frac{5}{8}.$$

We know that 5 and 8 cannot be divided evenly by any number.

$\frac{20}{32} = \frac{5}{8}$

$$\text{Therefore, } \frac{20}{32} = \frac{5}{8} \text{ in simplest form.}$$

PASS
WITH
CARE

Notice that in this example instead of dividing twice by 2 we could have divided by 4.

$$\text{That is, } \frac{20 \div 4}{32 \div 4} = \frac{5}{8}.$$

In order to do the problem quickly, divide by the *largest* common factor that we can find.

EXERCISES A.4

1. Reduce each of the following fractions to an equivalent fraction in simplest form (lowest terms):

   a) $\frac{4}{12}$ f) $\frac{15}{20}$

   b) $\frac{9}{27}$ g) $\frac{24}{40}$

   c) $\frac{6}{8}$ h) $\frac{30}{45}$

   d) $\frac{7}{15}$ i) $\frac{36}{54}$

   e) $\frac{12}{26}$ j) $\frac{21}{35}$

2. On a test, Bill answered 20 questions correctly out of 25.

   a)   What fractional part of the test did he answer correctly?

   b)   What fractional part of the test did he answer incorrectly?

3. The distance from Hoytesville to Canover is 63 miles. From Canover to Boontown is 42 miles. If you drive from Hoytesville to Canover, what fractional part of the distance from Hoytesville to Boontown have you travelled?

A.5 ADDITION AND SUBTRACTION OF FRACTIONS

Sometimes it is necessary to add or subtract fractions. Consider the following situation.

PROBLEM

George and Helen Tellman are a married couple whose Uncle Ted dies. They learn from Uncle Ted's lawyer that in his will George is to receive $\frac{3}{4}$ of Ted's ranch in Arizona and Helen receives $\frac{1}{6}$ of the ranch. George and Helen are wondering exactly what fraction of the whole ranch they have inherited together.

Solution:

In Section A.3 we saw that in order to add or subtract fractions we must keep *one basic rule* in mind:

   The denominators must be the same.

In our problem, George got $\frac{3}{4}$ of the ranch and Helen got $\frac{1}{6}$. In order to see what they received together $\frac{3}{4} + \frac{1}{6}$ must be added.

Finding a common denominator.

First both denominators must be made the same. Proceed as follows:

1. Multiply the denominators: $4 \times 6 = 24$.

2. Change each of the fractions $\frac{3}{4}$ and $\frac{1}{6}$ to equivalent fractions with denominators of 24 (see Section A.2).

$$\frac{3 \times 6}{4 \times 6} = \frac{18}{24};$$

$$\frac{1 \times 4}{6 \times 4} = \frac{4}{24}$$

$\frac{3}{4} = \frac{18}{24}$

$\frac{1}{6} = \frac{4}{24}$

3. Add the equivalent fractions by keeping the denominators the same and adding the numerators.

$$\frac{18}{24} + \frac{4}{24} = \frac{22}{24}$$

PASS
WITH
CARE

Therefore,

$$\frac{3}{4} + \frac{1}{6} = \frac{22}{24}$$

Can the fraction $\frac{22}{24}$ be reduced?

We see that both numbers have a factor of 2.

So,

2 divides 22 and 24 evenly.

$$\frac{22 \div 2}{24 \div 2} = \frac{11}{12}$$

$\frac{11}{12}$ of the ranch

$\frac{11}{12}$ is the fraction in simplest form. Why?

Therefore,

$$\frac{3}{4} + \frac{1}{6} = \frac{11}{12}.$$

So, George and Helen together inherited $\frac{11}{12}$ of Uncle Ted's ranch.

What happens if we have to *subtract* two fractions such as $\frac{3}{5} - \frac{1}{4}$?

The same *basic rule* applies as for addition:

The denominators must be the same.

Proceed as follows:

1. Multiply the denominators: $5 \times 4 = 20$.

2. Change each fraction, $\frac{3}{5}$ and $\frac{1}{4}$, into an equivalent fraction with denominator of 20.

$$\frac{3 \times 4}{5 \times 4} = \frac{12}{20};$$

$$\frac{1 \times 5}{4 \times 5} = \frac{5}{20}$$

Subtracting fractions

$$\frac{3}{5} = \frac{12}{20}$$

$$\frac{1}{4} = \frac{5}{20}$$

3. Subtract the equivalent fractions by keeping the denominators the same and subtracting the numerators.

$$\frac{12}{20} - \frac{5}{20} = \frac{7}{20}$$

Therefore, $\frac{3}{5} - \frac{1}{4} = \frac{7}{20}$.

Can $\frac{7}{20}$ be reduced? Why?

EXERCISES A.5

Perform the indicated operations in Exercises 1 to 12.

1. $\frac{1}{2} + \frac{1}{5}$

2. $\frac{1}{4} + \frac{2}{3}$

3. $\frac{2}{5} + \frac{3}{7}$

4. $\frac{1}{2} - \frac{1}{5}$

5. $\frac{3}{4} - \frac{2}{7}$

6. $\frac{7}{9} - \frac{3}{5}$

7. $\frac{5}{6} + \frac{7}{18}$

8. $\frac{3}{20} + \frac{5}{12}$

9. $\frac{8}{15} - \frac{3}{10}$

10. $\frac{1}{2} + \frac{1}{4} + \frac{1}{6}$

A common denominator can be found by multiplying the denominators together.
But is there an easier way?

11. $\frac{2}{3} + \frac{5}{12} + \frac{2}{9}$

12. $\frac{4}{5} + \frac{3}{10} - \frac{7}{20}$

13. What must be added to $\frac{4}{7}$ to make $\frac{2}{3}$?

14. Which of these two fractions is the larger number and how much larger? $\frac{4}{7}$ and $\frac{2}{3}$.

15. Jane is taking a poll and must interview a certain group of people. She interviews $\frac{1}{3}$ of the people before lunch and $\frac{2}{5}$ of the people after lunch.
    a) What fraction of the people has she interviewed?
    b) What fraction of the people must she still interview?

A.6 MULTIPLYING FRACTIONS

Joan wants to bake a cake. The recipe calls for $\frac{3}{4}$ of a cup of milk. However, Joan wants her cake to be only $\frac{1}{2}$ the size of the cake in the recipe. How much milk should she add?

What is $\frac{1}{2}$ of $\frac{3}{4}$?

  This problem illustrates a situation where it is necessary to multiply two fractions. Joan has to figure out what $\frac{1}{2}$ of $\frac{3}{4}$ of a cup of milk would be so that she can add that amount of milk. She must do the following:

$$\frac{1}{2} \times \frac{3}{4}$$

When multiplying two fractions we multiply the numerators together and the denominators together.

$$\text{So,} \frac{1}{2} \times \frac{3}{4} = \frac{1 \times 3}{2 \times 4} = \frac{3}{8}.$$

Therefore, Joan must add $\frac{3}{8}$ of a cup of milk in order to follow the recipe.

Why did we multiply in this way? In order to understand this we will consider a case involving a *fraction* and a *whole number*.

Amanda divided a pie into *four* equal parts. She gave one part ($\frac{1}{4}$) to Paul. He ate that piece quickly so she gave him another piece ($\frac{1}{4}$).

He also ate that piece so she finally gave him a third part of the pie ($\frac{1}{4}$). How much pie did Paul eat?

$$\frac{1}{4} + \frac{1}{4} + \frac{1}{4}$$

Paul got $\frac{1}{4}$ of the pie *three* times. We know that he got $\frac{3}{4}$ of the whole pie because

$$\frac{1}{4} + \frac{1}{4} + \frac{1}{4} = \frac{3}{4}.$$

But we see that we would arrive at the same conclusion by multiplying:

$$3 \times \frac{1}{4} \text{ (remember that } 3 = \frac{3}{1}\text{).}$$

$$3 = \frac{3}{1}$$

So, we see that $\frac{3}{1} \times \frac{1}{4} = \frac{3}{4}$ *if* we multiply the numerators and the denominators.

Here are two examples of multiplying fractions.

EXAMPLE 1

Multiply $\frac{2}{15}$ by 6.

$$6 = \frac{6}{1}$$

Solution:

$$\frac{6}{1} \times \frac{2}{15} = \frac{6 \times 2}{1 \times 15} = \frac{12}{15}$$

Notice that 12 and 15 have a common factor of 3.

3 divides 12 and 15 evenly.

$$\text{So, } \frac{12 \div 3}{15 \div 3} = \frac{4}{5}$$

Therefore, $6 \times \frac{2}{15} = \frac{12}{15} = \frac{4}{5}$

EXAMPLE 2

Multiply $\frac{2}{3}$ by $\frac{5}{6}$.

Solution:

$$\frac{2}{3} \times \frac{5}{6} = \frac{2 \times 5}{3 \times 6} = \frac{10}{18}$$

Can $\frac{10}{18}$ be reduced to simpler form?
We see that 2 is a common factor.

2 divides 10 and 18.

$$\text{So, } \frac{10 \div 2}{18 \div 2} = \frac{5}{9}$$

Therefore, $\frac{2}{3} \times \frac{5}{6} = \frac{10}{18} = \frac{5}{9}$.

EXERCISES A.6

Find the products:

1. $2 \times \frac{3}{7}$

2. $\frac{1}{8} \times 4$

3. $\frac{3}{7} \times \frac{2}{5}$

4. $\frac{3}{8} \times \frac{4}{5}$

5. $\frac{2}{3} \times \frac{6}{7}$

6. $\frac{3}{4} \times \frac{2}{5} \times 3$

7. $\frac{2}{9} \times \frac{3}{7} \times \frac{5}{6}$

8. $\frac{5}{11} \times \frac{1}{4} \times 0$

9. $\frac{1}{4} \times \frac{5}{12} \times \frac{2}{5}$

10. $\frac{4}{7} \times \frac{7}{8} \times \frac{2}{3}$

11. If chuck steak cost $1.40 a pound, how much will $\frac{3}{4}$ of a pound cost?

12. Tom spends $\frac{1}{5}$ of his income on clothes. Of this amount, he spends $\frac{2}{3}$ on suits. What fractional part of his income does he spend on suits?

A.7 DIVIDING FRACTIONS

In the study of probability we sometimes encounter what is known as a *complex fraction* (for example, when we consider *odds* in favor of an event). A complex fraction is one in which the numerator or the denominator, or both, involve fractions. For example,

Odds
p. 116

$$\frac{\frac{1}{2}}{\frac{3}{4}}$$

Complex fraction

is a complex fraction. How do we *simplify* such a fraction? In order to answer this question, we will first have to define a new term.

The reciprocal *of a fraction is another fraction which when multiplied by the first yields a product of 1.*

Reciprocal

For example, the reciprocal of $\frac{2}{3}$ is $\frac{3}{2}$ because

$$\frac{2}{3} \times \frac{3}{2} = \frac{6}{6} = 1.$$

The reciprocal of $\frac{4}{9}$ is $\frac{9}{4}$ because

$$\frac{4}{9} \times \frac{9}{4} = \frac{36}{36} = 1.$$

PASS
WITH
CARE

Note that the reciprocal of a fraction is obtained by *inverting* the original fraction.

The reciprocal of 5 is $\frac{1}{5}$

From Sec. A.2.

$\frac{3}{4} \times \frac{4}{3} = 1$

PASS
WITH
CARE

$\frac{1}{2} \div \frac{3}{4}$
is same as
$\frac{1}{2} \times \frac{4}{3}$

In the case of whole numbers, we know that the number 5, for example, can be written as $\frac{5}{1}$. So, the reciprocal of 5 is $\frac{1}{5}$.

That is, $\frac{5}{1} \times \frac{1}{5} = 1$.

Now, how do we simplify a complex fraction like

$$\frac{\frac{1}{2}}{\frac{3}{4}}$$

We know that we can multiply the numerator and denominator of a fraction by the same number and we get an equivalent fraction.

So, we multiply the numerator and the denominator by the *reciprocal of the denominator;* in this case, by $\frac{4}{3}$.

That is, $\dfrac{\frac{1}{2} \times \frac{4}{3}}{\frac{3}{4} \times \frac{4}{3}} = \dfrac{\frac{1}{2} \times \frac{4}{3}}{\frac{12}{12}} = \dfrac{\frac{1}{2} \times \frac{4}{3}}{1} = \frac{1}{2} \times \frac{4}{3}$.

Therefore, $\dfrac{\frac{1}{2}}{\frac{3}{4}} = \frac{1}{2} \times \frac{4}{3} = \frac{4}{6} = \frac{2}{3}$.

Note that this process shows that

$\dfrac{\frac{1}{2}}{\frac{3}{4}}$ is equivalent to $\frac{1}{2} \times \frac{4}{3}$.

We already know that

$\dfrac{\frac{1}{2}}{\frac{3}{4}}$ is the same as $\frac{1}{2} \div \frac{3}{4}$.

So, we now know that

$\frac{1}{2} \div \frac{3}{4}$ is the same as $\frac{1}{2} \times \frac{4}{3}$.

That is, if we want to divide one fraction by another we get the *same result* by *inverting the divisor* and *multiplying*.

In this case, we invert $\frac{3}{4}$ (to $\frac{4}{3}$) and multiply.

Here are some solved problems using this technique.

EXAMPLE 1

Simplify: $\dfrac{\frac{2}{5}}{\frac{6}{7}}$

Solution:

From above, we know that

$$\frac{\frac{2}{5}}{\frac{6}{7}} = \frac{2}{5} \div \frac{6}{7} = \frac{2}{5} \times \frac{7}{6}$$

Invert the divisor and multiply.

$$= \frac{14}{30} = \frac{7}{15}.$$

Therefore, $\quad \dfrac{\frac{2}{5}}{\frac{6}{7}} = \dfrac{7}{15}.$

EXAMPLE 2

Simplify:

$$\frac{3}{5} \div 6$$

Solution:

6 is equivalent to $\frac{6}{1}$.

$6 = \frac{6}{1}$

So, $\dfrac{3}{5} \div 6 = \dfrac{3}{5} \div \dfrac{6}{1} = \dfrac{3}{5} \times \dfrac{1}{6}$

Invert the divisor and multiply.

$$= \frac{3}{30} = \frac{1}{10}.$$

Therefore, $\dfrac{3}{5} \div 6 = \dfrac{1}{10}.$

EXAMPLE 3

Divide 8 by $\frac{2}{3}$.

Solution:

This means $8 \div \frac{2}{3}$.

$8 = \frac{8}{1}$

Invert the divisor and multiply.

$$8 \div \frac{2}{3} = \frac{8}{1} \div \frac{2}{3} = \frac{8}{1} \times \frac{3}{2}$$

$$= \frac{24}{2}$$

$$= 12$$

Therefore, $8 \div \frac{2}{3} = 12$.

EXERCISES A.7

Simplify each of the following fractions:

1. $\dfrac{\frac{1}{6}}{\frac{3}{5}}$

3. $\dfrac{\frac{2}{3}}{6}$

2. $\dfrac{\frac{2}{7}}{\frac{3}{7}}$

4. $\dfrac{8}{\frac{2}{5}}$

Find the quotient in each case:

5. $\frac{3}{4} \div \frac{6}{7}$

8. $\frac{3}{8} \div \frac{5}{6}$

6. $\frac{1}{5} \div \frac{4}{9}$

9. $\frac{3}{7} \div 6$

7. $\frac{2}{11} \div \frac{4}{7}$

10. $9 \div \frac{3}{5}$

11. How many hamburgers can be made from $\frac{3}{4}$ lb. of ground beef if each hamburger weighs $\frac{1}{8}$ of a pound?

12. A carpenter has a board 6 ft. long. He wishes to cut it into pieces, each 4 in. long. How many such pieces can he get from the board?

A.8 KINDS OF FRACTIONS

Up to this point we have been dealing with fractions such as $\frac{2}{3}, \frac{5}{7}, \frac{10}{12}, \frac{13}{19}$, etc., where the numerator is a smaller number than the denominator. Fractions of this type are usually called *proper fractions*.

    It is also possible to combine a whole number with a fraction: $5\frac{1}{2}$ pounds of sugar. (This number, of course, really means $5 + \frac{1}{2}$.) A number in this form is usually called a *mixed number* because we "mix" a whole number and a fraction to form the new number $5\frac{1}{2}$. A mixed number such as $5\frac{1}{2}$ can be expressed in another way—as a fraction.

    The way to express $5\frac{1}{2}$ as a fraction is as follows:

Proper fractions

$$5\frac{1}{2} = 5 + \frac{1}{2}$$

Mixed number

Express a mixed number as a fraction.

1. Multiply the whole number by the denominator:

$$5 \times 2 = 10$$

2. Add this product to the numerator and leave the denominator as it was:

$$\frac{10 + 1}{2} = \frac{11}{2}$$

      Therefore, $5\frac{1}{2} = \frac{11}{2}.$

Why is this method correct?

PASS
WITH
CARE

We must remember here that the line in a fraction can be thought of as a *division symbol*.

That is, for example,

$$\frac{8}{2} \text{ means the same as } 2\overline{)8}$$

So, $\frac{8}{2}$ = 4.

Therefore, for the example above, we have

$$\frac{11}{2} = \frac{10+1}{2} = \frac{10}{2} + \frac{1}{2} = 5 + \frac{1}{2} = 5\frac{1}{2} \; .$$

That is, $5\frac{1}{2} = \frac{11}{2}$.

Improper fraction

A fraction like $\frac{11}{2}$ where the numerator is larger than the denominator is called an *improper fraction*.

Such a fraction can always be represented by a mixed number.

Also, a mixed number can always be expressed as an improper fraction.

Note:

The idea of the line in a fraction representing the operation of division is very important because it is the basis for *changing a fraction into a decimal number,* as we will see in Appendix B.

Changing a fraction to a decimal.

Remember that when we use the division symbol ($\overline{)}$) we designate each part by a special name.

For example, in $3\overline{)15}^{5}$

Divisor
Dividend
Quotient

3 is called the *divisor*.
15 is called the *dividend*.
5 is called the *quotient*.

EXAMPLE 1

Change the mixed number $4\frac{2}{3}$ to an improper fraction.

Solution:

Follow these steps:

1. Multiply the whole number (4) by the denominator (3) and add the numerator (2).

$$4 \times 3 = 12,$$
$$12 + 2 = 14.$$

2. Make this result (14) the numerator and keep the same denominator (3).

$$\frac{14}{3}$$

Therefore, $4\frac{2}{3} = \frac{14}{3}$.

EXAMPLE 2

Convert $\frac{21}{5}$ to a mixed number.

Solution:

Divide 21 by 5 and the remainder becomes a numerator with the same denominator (5).

$$\begin{array}{r} 4 \\ 5\overline{)21} \\ 20 \\ \hline 1 \end{array}$$

5 divides 21 *four* times with a remainder of 1.

Therefore, $\frac{21}{5} = 4\frac{1}{5}$.

EXERCISES A.8

Express each of the following mixed numbers as an improper fraction.

1. $3\frac{1}{2}$

2. $5\frac{2}{7}$

3. $9\frac{3}{4}$

4. $10\frac{1}{5}$

5. $14\frac{4}{9}$

6. $21\frac{2}{3}$

7. $62\frac{5}{6}$

8. $101\frac{1}{4}$

Express each of the following improper fractions as a mixed number.

9. $\frac{18}{4}$

10. $\frac{26}{6}$

11. $\frac{32}{7}$

12. $\frac{44}{3}$

13. $\frac{54}{8}$

14. $\frac{76}{10}$

15. $\frac{91}{13}$

16. $\frac{141}{30}$

A.9 OPERATIONS ON MIXED NUMBERS

When adding, subtracting, multiplying, or dividing mixed numbers, we may proceed as follows:

Change the mixed numbers into improper fractions and perform the operation as in Sections A.5, A.6, and A.7.

This method will *always* work although in some cases it is not the easiest way to do the problem. Examine the following solved problems and then we will comment on a shorter method for some cases.

EXAMPLE 1

Simplify:

$$5\frac{2}{3} + 3\frac{1}{4}$$

Adding mixed numbers.

Solution:

1. Change to improper fractions.

$$5\frac{2}{3} = \frac{15+2}{3} = \frac{17}{3}$$

$$3\frac{1}{4} = \frac{12+1}{4} = \frac{13}{4}$$

Improper fractions.

2. Change $\frac{17}{3}$ and $\frac{13}{4}$ to equivalent fractions with a denominator of $3 \times 4 = 12$.

$$\frac{17 \times 4}{3 \times 4} = \frac{68}{12}$$

$$\frac{13 \times 3}{4 \times 3} = \frac{39}{12}$$

Equivalent fractions.

3. Add:

$$\frac{68}{12} + \frac{39}{12} = \frac{107}{12} = 8\frac{11}{12}$$

Therefore, $5\frac{2}{3} + 3\frac{1}{4} = 8\frac{11}{12}$.

EXAMPLE 2

Simplify:

$$7\frac{2}{5} - 4\frac{1}{3}$$

Subtracting mixed numbers.

Solution:

1. Change to improper fractions:

$$7\frac{2}{5} = \frac{35+2}{5} = \frac{37}{5}$$

$$4\frac{1}{3} = \frac{12+1}{3} = \frac{13}{3}$$

Improper fractions.

2. Change $\frac{37}{5}$ and $\frac{13}{3}$ to equivalent fractions with a denominator of $5 \times 3 = 15$.

Equivalent fractions.

$$\frac{37 \times 3}{5 \times 3} = \frac{111}{15}$$

$$\frac{13 \times 5}{3 \times 5} = \frac{65}{15}$$

3. Subtract:

$$\frac{111}{15} - \frac{65}{15} = \frac{46}{15} = 3\frac{1}{15}$$

Therefore, $7\frac{2}{5} - 4\frac{1}{3} = 3\frac{1}{15}$

EXAMPLE 3
Simplify:

$$3\frac{4}{5} \times 5\frac{1}{4}$$

Multiplying mixed numbers.

Solution:

1. Change to improper fractions.

$$3\frac{4}{5} = \frac{15+4}{5} = \frac{19}{5}$$

$$5\frac{1}{4} = \frac{20+1}{4} = \frac{21}{4}$$

2. Multiply numerators together and denominators together.

$$\frac{19}{5} \times \frac{21}{4} = \frac{399}{20} = 19\frac{19}{20}$$

Therefore, $3\frac{4}{5} \times 5\frac{1}{4} = 19\frac{19}{20}$

EXAMPLE 4

Simplify: $9\frac{3}{5} \div 5\frac{1}{3}$

Solution:

Dividing mixed numbers.

1. Change to improper fractions.

$$9\frac{3}{5} = \frac{45+3}{5} = \frac{48}{5}$$

$$5\frac{1}{3} = \frac{15+1}{3} = \frac{16}{3}$$

2. Divide. That is, invert $\frac{16}{3}$ and multiply.

$$\frac{48}{5} \div \frac{16}{3} = \frac{48}{5} \times \frac{3}{16} = \frac{144}{80}$$

$$= \frac{9}{5}$$

$$= 1\frac{4}{5}$$

Therefore, $9\frac{3}{5} \div 5\frac{1}{3} = 1\frac{4}{5}$.

In each of these four problems we first changed the mixed numbers into improper fractions and then proceeded as we did when dealing with proper fractions. We presented the problems in this way because this method *will always work.* However, at times it can become rather complicated.

This method *always* works.

There is another way to handle this kind of problem which can save some time—especially when the whole numbers become larger. The following problem illustrates this other approach.

Another method of combining mixed numbers.

EXAMPLE 5
Simplify:

$$25\frac{2}{3} + 34\frac{3}{5}$$

Solution:

Adding mixed numbers.

1. Combine the whole numbers first. That is,

$$25 + 34 = 59.$$

2. Combine the fractions next. That is,

$$\frac{2}{3} + \frac{3}{5}$$

$$\frac{2}{3} + \frac{3}{5} = \frac{10}{15} + \frac{9}{15} = \frac{19}{15} = 1\frac{4}{15}.$$

3. Combine the two results. That is,

$$59 + 1\frac{4}{15} = 60\frac{4}{15}.$$

$$\text{Therefore, } 25\frac{2}{3} + 34\frac{3}{5} = 60\frac{4}{15}.$$

Do this same problem by the previous method and determine which you prefer.

Addition or Multiplication. This second method of handling problems that involve mixed numbers is often easier when the problems involve the operations of *addition* or *multiplication*. Problems involving subtraction or division are sometimes complicated by the second method.

Try some of the following exercises using both methods and decide for yourself which method you prefer to use.

EXERCISES A.9

Perform the indicated operations.

1. $4\frac{2}{3} + \frac{4}{5}$
2. $6\frac{1}{2} + 3\frac{5}{6}$
3. $13\frac{3}{4} + 8\frac{2}{7}$
4. $34\frac{1}{6} + 21\frac{3}{5}$
5. $6\frac{2}{3} - 2\frac{1}{2}$
6. $9\frac{2}{5} - 5\frac{3}{4}$
7. $17\frac{1}{2} - 11\frac{3}{7}$
8. $36 - 15\frac{2}{3}$
9. $\frac{3}{4} \times 4\frac{1}{2}$
10. $5\frac{1}{3} \times 7\frac{2}{5}$
11. $9\frac{5}{6} \times 14\frac{2}{3}$
12. $21\frac{2}{7} \times 32\frac{3}{5}$
13. $1\frac{2}{3} \div 6\frac{2}{3}$
14. $6\frac{3}{4} \div 9\frac{1}{3}$
15. $11\frac{1}{4} \div 3\frac{3}{8}$
16. $15\frac{2}{5} \div 6\frac{1}{4}$

17. A two-piece dress requires $2\frac{1}{3}$ yards of material for one piece and $3\frac{3}{4}$ yards for the other. How much material is needed for the dress?

18. A nail that is $4\frac{1}{3}$ inches long is driven through a piece of wood that is $2\frac{3}{4}$ inches thick. How much of the nail will stick out the other side?

19. At a cruising speed of $18\frac{1}{2}$ m.p.h., how far will a boat travel in $4\frac{1}{4}$ hours?

20. How many pieces of rope, each $1\frac{3}{4}$ feet long, can be cut from a length of rope $10\frac{1}{2}$ ft. long?

Appendix B
Decimals

B.1 WHAT IS A DECIMAL?

Our number system is a decimal system because it contains *10* digits:

0, 1, 2, 3, 4, 5, 6, 7, 8, 9.

The system uses a point (usually called a *decimal point*) to distinguish whole numbers from numbers that are *less than 1*. For example, a number such as 516.38 means

516 + .38

where 516 is a whole number and .38 is a decimal number that is less than 1.

The following example illustrates the name that is given to each of the places in a number. For the number 4278.613, we have the following:

thousands	hundreds	tens	units		tenths	hundredths	thousandths
4	2	7	8	·	6	1	3

In our decimal system

.6 means $\frac{6}{10}$ (six *tenths*)

.61 means $\frac{61}{100}$ (sixty-one *hundredths*)

.613 means $\frac{613}{1000}$ (six hundred thirteen *thousandths*)

Keep in mind at all times that a *proper fraction* or a *decimal* always represent a number that is *less than 1*.

Decimal system = 10 digits

.38 < 1

$.6 \ = \frac{6}{10}$

$.61 \ = \frac{61}{100}$

$.613 \ = \frac{613}{1000}$

An important point.

195

That is, for example,

$$\frac{2}{5} < 1$$

$$.65 < 1$$

$$\frac{2}{5} < 1$$

$$.65 < 1$$

Fractions and decimals are the two methods most commonly employed to express numbers that are less than 1. It is necessary to be able to convert a decimal to a fraction and vice versa.

PASS
WITH
CARE

Note:

Since whole numbers are written to the *left* of the decimal point, it is understood that a whole number has a point to its *right*.

For example,

328 is the same as 328.00

EXERCISES B.1

Write the name for each of the numbers in exercises 1 through 12 as in the examples given:

5.69 = five and sixty nine hundredths

14.8 = fourteen and eight tenths

1.   .6
2.   .29
3.   .08
4.   .80
5.   .008
6.   8.0
7.   .507
8.   9.41
9.   312.8
10.   76.12
11.   11.262
12.   108.50
13.   Which of the numbers in Exercises 1 to 9 are less than 1?

B.2 CONVERTING A DECIMAL NUMBER TO A FRACTION

As we noted in Section B.1, in our decimal number system

.7 means $\frac{7}{10}$ (seven *tenths*)

.23 means $\frac{23}{100}$ (twenty three *hundredths*)

.806 means $\frac{806}{1000}$ (eight hundred six *thousandths*)

and so on.

$$.7 = \frac{7}{10}$$
$$.23 = \frac{23}{100}$$
$$.806 = \frac{806}{1000}$$

Therefore, to change a decimal to a fraction you simply do the following:

1. Count the number of digits after the decimal point. The *denominator* will have that many zeroes following the 1.

2. The number after the point becomes the *numerator*.

Thus, for example:

$$.67 = \frac{67}{100}$$
$$.9 = \frac{9}{10}$$

Note that *zeroes* can be added at the end of a decimal without changing the value of the number.

For example:

$$.4 = .40 = .400$$

Zeroes can be added at the end of a decimal.

This is true because when we convert these decimals to fractions, we see that they are equivalent.

$$.4 = \frac{4}{10} = \frac{4 \div 2}{10 \div 2} = \frac{2}{5}$$
$$.40 = \frac{40}{100} = \frac{40 \div 20}{100 \div 20} = \frac{2}{5}$$
$$.400 = \frac{400}{1000} = \frac{400 \div 200}{1000 \div 200} = \frac{2}{5}$$

EXERCISES B.2

Convert each of the following decimals to the equivalent fraction in simplest form:

1. .8
2. .42
3. .875
4. .06
5. .57

6. 3.6
7. 23.75
8. .045
9. .0045
10. 721.9

B.3 CONVERTING A FRACTION TO A DECIMAL NUMBER

As discussed in Section B.1, proper fractions and decimals are the means used to represent numbers that are less than 1.

We can change any fraction to its decimal equivalent by applying an idea that we saw in Section A.7. That is,

The line in a fraction is a division *symbol.*

$\dfrac{2}{5}$ becomes $5\overline{)2}$.

Therefore, $\dfrac{2}{5}$ means "divide 5 into 2" or "2 divided by 5 "

or $5\overline{)2}$

$\dfrac{16}{45}$ means "divide 45 into 16" or "16 divided by 45"

or $45\overline{)16}$

So, in order to change a fraction into a decimal we divide the numerator by the denominator. (Remember that for a whole number the decimal point is at the right of the number).

In these problems we will also review the process of *division.*

EXAMPLE 1

Convert $\frac{2}{5}$ to an equivalent decimal.

Solution:

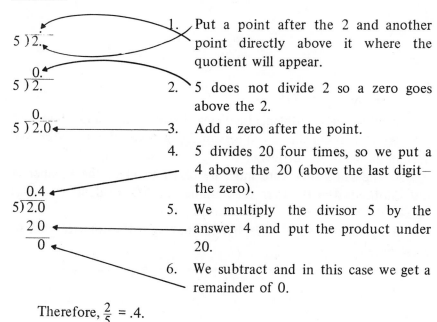

1. Put a point after the 2 and another point directly above it where the quotient will appear.

2. 5 does not divide 2 so a zero goes above the 2.

3. Add a zero after the point.

4. 5 divides 20 four times, so we put a 4 above the 20 (above the last digit—the zero).

5. We multiply the divisor 5 by the answer 4 and put the product under 20.

6. We subtract and in this case we get a remainder of 0.

Therefore, $\frac{2}{5}$ = .4.

EXAMPLE 2

Convert $\frac{3}{8}$ to an equivalent decimal.

In this problem the steps are the same as in Problem 1. We continue to add zeroes after the decimal point until we get a remainder of zero.

Examine carefully. This method is important.

Solution:

$$
\begin{array}{r}
0.375 \\
8\overline{)3.000} \\
\underline{2\,4} \\
60 \\
\underline{56} \\
40 \\
\underline{40} \\
0
\end{array}
$$

Therefore, $\frac{3}{8}$ = .375.

In all division problems involving decimals, it is important that the decimal point in the quotient be placed *directly over* the decimal point in the dividend.

At the same time care must be taken to place the number in the quotient directly over the number in the dividend that was "taken down" to make the division possible.

Many *mistakes* are made by not paying sufficient attention to these details.

PASS
WITH
CARE

EXERCISES B.3

Convert each of the following fractions to a decimal equivalent:

1. $\frac{1}{2}$

2. $\frac{1}{4}$

3. $\frac{3}{4}$ (remember that $\frac{3}{4} = 3 \times \frac{1}{4}$)

4. $\frac{1}{5}$

5. $\frac{3}{5}$

6. $\frac{1}{8}$

7. $\frac{5}{8}$

8. $\frac{1}{10}$

9. $\frac{1}{20}$

10. $\frac{3}{20}$

B.4.   APPROXIMATING A DECIMAL NUMBER (ROUNDING OFF)

In Section B.3, we converted the fractions $\frac{2}{5}$ and $\frac{3}{8}$ to decimal numbers. In each case, the division process eventually gave us a remainder of 0—and so the process ended. We found that

$$\frac{2}{5} = .4$$

$$\frac{3}{8} = .375$$

Decimal equivalents.

In some cases we find that instead of continuing the division process until we get a remainder of 0, it is convenient to stop the process at some point and give an *approximate* decimal equivalent for the fraction instead of an *exact* one. (In some cases, we *must* give an approximate answer because the division process *never ends.* Converting $\frac{1}{3}$ to a decimal is an example of such a case.)

Approximate decimal equivalent.

We never get a remainder of 0.

Let us use the conversion of $\frac{1}{3}$ to a decimal as an example of approximating a decimal number.

EXAMPLE 1

Write $\frac{1}{3}$ as a decimal number.

Solution:

$$
\begin{array}{r}
.333 \\
3\overline{)1.000} \\
\underline{9} \\
10 \\
\underline{9} \\
10 \\
\underline{9} \\
1
\end{array}
$$

We see that we will continue to get more 3's in the quotient if we continue to divide. The process will never end. Let us write $\frac{1}{3}$ as a decimal to *two places.* This means that we write $\frac{1}{3}$ as a decimal having just *two digits* to the right of the point.

Approximate (round off) to two places.

PASS
WITH
CARE

We know $\frac{1}{3}$ = .3333...

How to approximate.

We want a decimal to *two places*.

1. Look at the *next* digit (in this case the *third* digit after the point).
2. If this digit is 5 or greater, we increase the second digit by one.
3. If this digit is 4 or less, we leave the second digit as it is.

   In this case, 3 < 5.

Therefore, $\frac{1}{3}$ ≈ .33 to *two places. The symbol ≈ means "approximately equal to."*

EXAMPLE 2

a) Write 2.134 as a decimal to *two places*

b) Write .081 as a decimal to *one place.*

Solution:

a) Look at the *third* digit after the point.
    It is 4, and 4 < 5.
    So, 2.13④ ≈ 2.13 (to *two* places).

b) Look at the *second* digit after the point.
    It is 8, and 8 > 5.
    So we add one to the digit 0, .0⑧1 ≈ .1 (to *one* place).

Other terms are often used when we are asked to approximate decimals.

Since, for example,

$$.4 = \frac{4}{10} \quad \text{(four tenths)}$$

Tenths place.

the first digit after the point is said to be in the *"tenths place"* (as noted in Section B.1).

Since, for example,

$$.32 = \frac{32}{100} \quad \text{(thirty two hundredths)}$$

Hundredths place.

The second digit after the point is said to be in the *"hundredths place."* The third digit after the point is said to be in the

Thousandths place.

thousandths place, and so on.

For example, look at the decimal .2064

The digit 2 is in the *tenths* place.
The digit 0 is in the *hundredths* place.
The digit 6 is in the *thousandths* place.
The digit 4 is in the *ten thousandths* place.

PASS
WITH
CARE

EXAMPLE 3

Write the decimals 23.14 and 6.271 to the *nearest tenth.*

Nearest tenth

Solution:

"To the nearest tenth" means the same as "to one decimal place."
That is, leave only *one* digit to the right of the decimal point.

For 23.14, the second digit, 4, is less than 5, so

$4 < 5$

$$23.1④ \approx 23.1 \text{ (nearest tenth)}.$$

For 6.271, the second digit, 7, is greater than 5,

$7 > 5$

$$\text{so } 6.2⑦1 \approx 6.3 \text{ (nearest tenth)}$$

Notice that

$$6.27① \approx 6.27 \text{ (nearest } hundredth\text{). Why?}$$

Nearest hundredth.

EXERCISES B.4

In Exercises 1 through 9, approximate each decimal to the *nearest hundredth,* to the *nearest tenth,* and to the nearest *whole number.*

1. 6.238
2. .7516
3. 79.613
4. 274.397
5. .0652
6. 14.083
7. 5.347
8. 44.491

9. Complete the following table. Where you must approximate, give the answer to the nearest *thousandth*. (See Exercises B 3.)

Fraction	$\frac{1}{2}$	$\frac{1}{3}$	$\frac{2}{3}$	$\frac{1}{4}$	$\frac{3}{4}$	$\frac{1}{5}$	$\frac{2}{5}$	$\frac{3}{5}$	$\frac{4}{5}$	$\frac{1}{6}$	$\frac{5}{6}$	$\frac{1}{7}$	$\frac{3}{7}$	$\frac{1}{8}$	$\frac{3}{8}$	$\frac{1}{9}$	$\frac{5}{9}$	$\frac{1}{10}$	$\frac{1}{12}$
Decimal																			

B.5 ADDITION AND SUBTRACTION OF DECIMAL NUMBERS

Addition and subtraction of decimals is similar to addition and subtraction of whole numbers. We must, however, be careful to follow one important rule:

When adding or subtracting decimals, the decimal points must be kept one under the other.

Failure to line up the points carefully when adding or subtracting decimals is a major cause of mistakes. *Be careful!*

Here are some solved problems to illustrate this point.

EXAMPLE 1

Adding decimals

Add: .621, 24.2, 7.3, 159

Solution:

Line up the points

```
  .621
 24.2
  7.3
159.
───────
191.121
```

Line up the numbers according to the points. The decimal point in the sum is directly under the others.

EXAMPLE 2

Subtract .62 from 25.5

Solution:

$$\begin{array}{r} 25.50 \\ .62 \\ \hline 24.88 \end{array}$$

Remember that adding a zero to the end of a decimal does not change the number.

Line up the points

EXERCISES B.5

Add:

1. 47.34 + .86 + 86.89 + 3.71
2. 3.742 + 13.14 + .83 + 26.4
3. .602 + 9.7 + 263.79 + 15.6
4. 7.41 + .063 + 37.8 + 26
5. 29.2 + 17 + .80 + 4.751

Subtract:

6. 59.74 -7.41
7. 6.324 -.52
8. 15.213 -7.646
9. 275.7 -40.33
10. 3.46 -.082
11. Six runners clocked the following times for the hundred-yard dash: 9.8, 10.4, 11.1, 10.2, 10.9, 9.7 seconds. If the four fastest runners ran the same times in a 400-yard relay race, what would be the best time for the relay race?
12. A man owns 143.62 acres of land. He sells three parcels of land of the following sizes: 14.34, 47.5, and 9.71 acres. How much land does the man still own?

B.6   MULTIPLICATION OF DECIMAL NUMBERS

When two decimal numbers are multiplied the process is the same as for whole numbers—*except* that care must be taken to position the decimal point correctly in the product.

For example, multiply 64.23 by 2.6.

Position the numbers as usual and do the multiplication without worrying about the decimal point *until the end* of the multiplication process. (It is not necessary to put the points directly under each other as for addition and subtraction.)

Multiplying decimals

Where does the decimal point go in the product?

$$
\begin{array}{r}
64.23 \\
\times\ 2.6 \\
\hline
38538 \\
12846 \\
\hline
166998
\end{array}
$$

Multiply as for whole numbers.

Do not worry about the decimal point until the end.

To position the decimal point correctly in the product proceed as follows:

1. Count the number of places after the decimal point in each of the two numbers being multiplied.
   64.23 has *two* places after the point.
   2.6 has *one* place after the point.
2. Add these places together: 2 + 1 = 3.
3. The decimal point is situated in the product so that there are *3* places after the point.

Therefore, 64.23 × 2.6 = 166.998.

EXAMPLE

Multiply 6.31 by .742

$$
\begin{array}{r}
6.31 \\
\times\ .742 \\
\hline
1262 \\
2524 \\
4417 \\
\hline
4.68202
\end{array}
$$

1. Multiply as for whole numbers.

2. Count the number of decimal places being multiplied. (The number is 5.)   5 decimal places

3. So, there must be 5 places after the decimal point in the product.   5 digits after the decimal point

Therefore, 6.31 $\times$ .742 = 4.68202.

What would the answer be to the *nearest tenth*?   Nearest tenth

What would the answer be to the nearest *hundredth*?   Nearest hundredth

EXERCISES B.6

Multiply:

1. 526 $\times$ 7.4
2. 7.321 $\times$ 25.6
3. .645 $\times$ 2.31
4. 80.3 $\times$ .9
5. 43.061 $\times$ .04
6. 157.4 $\times$ 20.8
7. .072 $\times$ 16.4
8. 2.006 $\times$ .035
9. A racing car travelled at a steady speed of 142.26 miles per hour for .74 hours. How far did it travel in this time?
10. Water weighs 62.5 lb. per cubic foot. Cork weighs .24 times the weight of water.
    a) Does cork weigh more or less than water?
    b) What does a cubic foot of cork weigh?

B.7   DIVISION OF DECIMAL NUMBERS

In considering division of decimals, we will take a look at two cases:

1. Division of a decimal by a whole number.
2. Division of a decimal by a decimal.

CASE 1 EXAMPLE

Divide 3.912 by 24

$$\begin{array}{r} 0. \\ 24\overline{)\,3.912} \end{array}$$

1. Place a decimal point in the quotient directly above the point in the *dividend*.

2. 24 does not divide 3. Place a 0 above the digit 3.

$$\begin{array}{r} 0.1 \\ 24\overline{)\,3.912} \\ 2\ 4 \\ \hline 1\ 5 \end{array}$$

3. 24 divides 39 *once*. Place a 1 above 39 (over the digit 9).

4. Multiply 24 by the 1 and put the product under the 39.

5. Subtract—we get 15.

$$\begin{array}{r} 0.16 \\ 24\overline{)\,3.912} \\ 2\ 4 \\ \hline 1\ 51 \\ 1\ 44 \\ \hline 7 \end{array}$$

6. "Bring down" the next digit (the digit 1) and place it after the 15.

7. 24 divides 151 *six* times. Place the 6 above the digit 1 that we brought down.

8. Multiply 24 by 6 and put the product under 151.

9. Subtract—we get 7.

10. Bring down the digit 2.

11. 24 divides 72 *three* times. Place the 3 above the digit 2.

12. Multiply 24 by 3 and put the product under 72.

13. Subtract—we get 0.

14. The division is complete.

```
       0.163
   24)3.912
       2 4
       1 51
       1 44
         72
         72
          0
```

In this case we got a remainder of 0. This does not always happen.

Therefore, 3.912 ÷ 24 = 0.163.

Note:

We can check to see if the answer is correct. Multiply the quotient *(.163) by the* divisor *(24) and we should get the* dividend *(3.912) as the answer. This multiplication works as a check on our division because multiplication is the* inverse operation *of division. This means simply that one operation is the reverse of the other. We could also check a multiplication problem by an appropriate division.*

This method is a check because multiplication is the inverse operation to division.

CASE 2 EXAMPLE

Divide 6.241 by .41 and give the answer to the nearest tenth.

Solution:

To divide by a decimal, first determine where the decimal point will be in the quotient. Then proceed as before.

Examine these steps carefully.

1. There are *two* digits after the decimal point in the divisor.

$$.41\overline{)6.241}$$

2. Therefore, move the decimal point *two* places to the right in the divisor *and* also in the dividend. (See the note at the end of the problem for *why* this does not change the problem.)

$$41\overline{)624.1}$$

3. We now have the problem as it appears on the left.

4. We proceed with the division as we did in Case 1.

We do not get a remainder of 0.

$$\begin{array}{r} 15.2 \\ 41\overline{)624.1} \\ 41 \\ \hline 214 \\ 205 \\ \hline 91 \\ 82 \\ \hline 9 \end{array}$$

5. In doing the division, we do not get a remainder of 0 at any time. *Then when do we stop?*

6. The problem asked for the answer to the nearest tenth (or *one* decimal place). We continue until we get *one more decimal place* than asked for. (We do this so that we can approximate or "round off" our decimal answer properly.)

Approximate the answer.

$$\begin{array}{r} 15.22 \\ 41\overline{)624.10} \\ 41 \\ \hline 214 \\ 205 \\ \hline 91 \\ 82 \\ \hline 90 \\ 82 \\ \hline 8 \end{array}$$

7. Add a zero after the last digit and "bring it down."

8. Divide 41 into 90.

9. We now have a quotient with *two* decimal places.
Since 2 is not greater than 5, we get 15.2.

Therefore, $6.241 \div .41 = 15.2$ to the nearest tenth.

Remember that this is an *approximate* answer. If you now check the division by doing the following:

$$15.2$$
$$\times .41$$

An approximate answer—not an *exact* answer. (Rounding error).

you find that you do not get 6.241 *exactly*. You get a number that is very close to it. Try this.

Note:

Why is it that if we move the decimal points in both the divisor and the dividend the problem remains the same? Remember that

$$.41\overline{)6.241} \text{ is the same as } \frac{6.241}{.41}$$
$$\text{and } \frac{6.241 \times 100}{.41 \times 100} = \frac{624.1}{41}.$$

By multiplying the numerator and denominator by 100 we get an equivalent fraction. We also get a whole number (41) as the denominator or divisor—which is what we want.

EXERCISES B.7

Divide and check your answer by multiplication:
1. $655.62 \div 42$
2. $11.712 \div .61$
3. $1.8432 \div .072$

Divide and give your answer to the nearest *hundredth*:
4. $47.28 \div 3.7$
5. $.83216 \div .301$
6. $.79003 \div 52.1$
7. $807.9 \div .502$
8. $.04638 \div 27.1$
9. Carl drove 607.5 miles on a business trip. His car averaged 16.2 miles to a gallon of gasoline. How many gallons of gasoline did he use?
10. A truck can carry 6.7 tons of earth in one load. How many trips will be necessary to move 107.2 tons of earth?

Appendix C
Percents

C.1 WHAT IS A PERCENT?

We see the word "percent" used so often and in what seem to be so many different ways that we may forget that the word has a very simple meaning.

The word "cent" comes from the Latin word meaning "hundred." The term percent simply means "per hundred."

Per hundred.

For example, if we know that 20% of a particular group of people have cavities in their teeth, this is the same as saying that "20 per 100" of the people have cavities. Meaning that out of every 100 people in the group, 20 have cavities.

20 per 100

Since 20% means 20 per 100, it has the same meaning as $\frac{20}{100}$.

$20\% = \frac{20}{100}$

What we are saying is that the use of the percent symbol (%) is just *another way* of denoting a fraction with a denominator of 100. (Think of the symbol % as meaning $\frac{}{100}$.) For example,

$$36\% \text{ is another way of writing } \frac{36}{100}.$$

$$12\tfrac{1}{2}\% \text{ is another way of writing } \frac{12\frac{1}{2}}{100}.$$

$$246\% = \frac{246}{100}$$

PASS
WITH
CARE

EXAMPLE 1

Carol has a coin collection containing 800 coins. 15% of the coins are dimes. How many of the coins are dimes?

Solution:

15% means that 15 out of every 100 coins are dimes. There are 800 coins, so there are 8 groups of 100 coins. In each group there are 15 dimes. So, 8 × 15 = 90. Therefore, there are 90 dimes in Carol's collection.

15 per 100

EXAMPLE 2

Phil borrows $500 from a bank. At the end of one year, Phil will owe the bank 12% interest besides the money he borrowed. How much money will he owe in interest?

Solution:

$12 per $100

12% interest means $12 per $100. That is, Phil has to pay $12 for every $100 he borrowed. He borrowed

$$\$500 \text{ (or } 5 \times \$100).$$

So, he has to pay

$$5 \times \$12 = \$60.$$

Therefore, Phil will owe $60 in interest after one year.

EXERCISES C.1

In these problems, use only the definition of percent as meaning "per hundred."

1. Lucille had 100 bars of candy to sell. On Monday she sold 23% of them. How many did she sell?

2. Peter had 200 bars of candy to sell. On Monday he sold 18% of them. How many did he sell?

3. A farmer owns 300 acres of land. He plows 60 acres on a particular day. What percent of his land did he plow?

C.2 EXPRESSING A NUMBER AS A PERCENT

We are often asked to express a number as a percent. The principle to remember in all such cases is:

A percent is the same as a fraction with denominator of 100.

For example,

$$18\% = \frac{18}{100}$$

$$\frac{5}{100} = 5\%$$

That is, 18% and $\frac{18}{100}$ both represent the same number.

So, if we are asked to express any number as a percent, we just *write that number as an equivalent fraction* with a *denominator of 100.*

Let us look at some problems.

EXAMPLE 1

Express $\frac{3}{4}$ as a percent.

Express a fraction as a percent.

Solution:

We simply express $\frac{3}{4}$ as an equivalent fraction with a denominator of 100. We see that we must multiply the denominator 4 by 25 to get 100.

$$\text{So, } \frac{3 \times 25}{4 \times 25} = \frac{75}{100} = 75\%.$$

$$\frac{3}{4} = \frac{75}{100}$$

Therefore, $\frac{3}{4} = 75\%$.

Any fraction *can be expressed as a percent—by changing it to an equivalent fraction with a denominator of 100.*

EXAMPLE 2

Express a decimal as a percent.

Express .6 as a percent.

Solution:

We know that .6 means $\frac{6}{10}$ (six tenths). So we convert $\frac{6}{10}$ to an equivalent fraction with a denominator of 100.

$$\frac{6 \times 10}{10 \times 10} = \frac{60}{100} = 60\%.$$

Therefore, $.6 = \frac{6}{10} = 60\%$.

A decimal can be changed to a percent by expressing it as a fraction first (with a denominator of 100).

$6 = \frac{6}{10} = \frac{60}{100}$

It is important to note that a decimal which has *two places*— for example, .38 — is changed to a percent by simply removing the point and adding the percent symbol. So, .38=38%. This is so because

.38 = 38%

$$.38 = \frac{38}{100} = 38\%.$$

In similar fashion,

$$.21 = \frac{21}{100} = 21\%$$

.08 = 8%

$$.08 = \frac{8}{100} = 8\%$$

$$3.47 = \frac{347}{100} = 347\%.$$

EXAMPLE 3

Express 3 as a percent.

Solution:

How do we express a whole number as a percent? We have to remember that $3 = \frac{3}{1}$. That is, any whole number can be expressed as a fraction with a denominator of 1. Now, change $\frac{3}{1}$ to an equivalent fraction with denominator of 100.

$$\frac{3 \times 100}{1 \times 100} = \frac{300}{100} = 300\%$$

Therefore, $3 = \frac{3}{1} = 300\%$.

At this time, we should pause to examine an interesting point.

We have just shown that in order to express a fraction as a percent, we convert it into an equivalent fraction with denominator of 100.

But this is not always the *easiest* way to do it. It is often easier to change the fraction into a decimal.

For example, change $\frac{11}{15}$ to a percent. In this case, it is easier to change the fraction to a decimal to *two places* (nearest hundredth).

$$
\begin{array}{r}
.733 \\
15\,)\overline{11.000} \\
\underline{10\ 5} \\
50 \\
\underline{45} \\
50 \\
\underline{45} \\
5
\end{array}
$$

Therefore, $\frac{11}{15} \approx .73 \approx 73\%$.

Express a whole number as a percent.

$3 = \dfrac{3}{1}$

$3 = \dfrac{300}{100}$

EXERCISES C.2

Convert the decimals in Exercises 1 through 10 to equivalent percents:

1.	.37	6.	1.42
2.	.54	7.	4.0
3.	.4	8.	.0053
4.	.04	9.	7.5
5.	.038	10.	.02

Convert the fractions in Exercises 11 through 20 to equivalent percents. (If it is necessary to approximate, give answer to nearest *whole percent*.)

11.	$\frac{2}{5}$	16.	$\frac{3}{7}$
12.	$\frac{3}{8}$	17.	$\frac{5}{12}$
13.	$\frac{2}{3}$	18.	$\frac{7}{5}$
14.	$\frac{5}{4}$	19.	$\frac{16}{25}$
15.	$\frac{2}{9}$	20.	$\frac{12}{200}$

C.3   EXPRESSING A PERCENT AS A FRACTION OR A DECIMAL

Expressing a percent as a fraction is a very simple procedure.

In Section C.1. we noted that

A percent is the same as a fraction with a denominator of 100.

Think of % as meaning $\frac{}{100}$. So, for example, 18% means $\frac{18}{100}$.

EXAMPLE 1

Express 54% as a fraction.

Express a percent as a fraction.

Solution:

$$54\% \text{ means } \frac{54}{100}.$$

We can reduce this fraction by dividing by 2 (a common factor).

$$\frac{54 \div 2}{100 \div 2} = \frac{27}{50}$$

Therefore, $54\% = \frac{54}{100} = \frac{27}{50}$.

Expressing a percent as a decimal is just as simple.

EXAMPLE 2

Express 42% as a decimal.

Express a percent as a decimal.

Solution:

$$42\% = \frac{42}{100}.$$

Change the fraction to a decimal by dividing.

$$
\begin{array}{r}
.42 \\
100\overline{)42.00} \\
40\ 0 \\
\hline
2\ 00 \\
2\ 00 \\
\hline
0
\end{array}
$$

Therefore, $42\% = \frac{42}{100} = .42$.

$42\% = .42$

Note a very interesting point that this example illustrates:

A percent is changed to a decimal by dropping the percent symbol (%) and moving the decimal point two *places to the left.*

PASS
WITH
CARE

So, for example,

$$34\% = \frac{34}{100} = .34$$

$$6\% = \frac{6}{100} = .06$$

$$271\% = \frac{271}{100} = 2.71.$$

EXERCISES C.3

Convert these percents to equivalent fractions in lowest terms:

1. 30%
2. 42%
3. 8%
4. 125%
5. 12.5%

6. 1.25%
7. .8%
8. 62.5%
9. 72%
10. 350%

11 to 20. Convert the percents in Exercises 1 to 10 to equivalent decimals.

C.4 PROBLEMS INVOLVING PERCENTS

Most problems involving percents deal with *comparisons* of two numbers. One number is compared *to* a second number which is called the *base*.

Use of percent:
compare two numbers.

For example, Bill buys a boat for $100 and sells it for $120.

If we ask how much profit he made—the answer is $20.

But we could ask what *percent* profit did he make. This means that we *compare* his profit ($20) *to* his cost ($100). The cost is the *base*.

Compare profit to cost.

We express this as a fraction $\frac{20}{100}$ (the base becomes the denominator.)

$$\text{We know } \frac{20}{100} = 20\%$$

So, Bill made a profit of 20% on selling the boat.

20% profit.

This is a simple example of the way in which percent is normally used.

In the following sections we will examine some of the common problems involving the use of percent—and how to solve them.

C.5 FIND THE PERCENT ONE NUMBER IS OF ANOTHER

In problems of this type, we must determine which number is the *base* for the comparison.

EXAMPLE 1

4 is what percent *of* 16?

Solution:

We are asked to *compare* 4 to 16 where 16 is the *base*. So, 16 becomes the denominator of the fraction.

Compare 4 to 16.

$$\frac{4}{16}$$

$$\text{We get } \frac{4}{16} = \frac{1}{4} = .25 = 25\%.$$

Therefore, 4 is 25% of 16.

EXAMPLE 2

What percent *of* 42 is 28?

Solution:

In this case, 42 is the *base*.

Compare 28 to 42.

$$\frac{28}{42}$$

$$\text{So, } \frac{28}{42} = \frac{2}{3} \approx .67 = 67\%.$$

Therefore, 28 is approximately 67% of 42.

EXAMPLE 3

What percent *of* 30 is 75?

Solution:

Be careful here. 30 is the *base* and the larger number 75 is compared to it.

Compare 75 to 30.

$$\frac{75}{30}$$

$$\text{So, } \frac{75}{30} = \frac{5}{2} = 2.5 = 250\%.$$

Therefore, 75 is 250% of 30.

EXAMPLE 4

A ranch owner has 24 horses. He sells 6 to another rancher.

    a)   What percent of the horses did he sell?

    b)   What percent of the horses did he keep?

Solution:

    a)   He sold 6 *of* the 24 horses. 24 is the base.

Compare 6 to 24.

$$\frac{6}{24}$$

$$\text{So, } \frac{6}{24} = \frac{1}{4} = .25 = 25\%$$

Therefore, he sold 25% of the horses.

    b)   He kept 18 of the 24 horses.

$$\text{So, } \frac{18}{24} = \frac{3}{4} = .75 = 75\%.$$

Therefore, he kept 75% of the horses.

Notice that 25% + 75% = 100%.

We could have found the percent he kept by subtracting the percent he sold from 100%.

That is, 100% – 25% = 75%.

This is so because the 24 horses represented the whole herd or *100%* of the rancher's horses. So, we just subtract the percent he sold (25%) from the total (100%) to find the percent he kept (75%).

EXERCISES C.5

1. 18 is what percent of 72?

2. 6 is what percent of 15?

3. 25 is what percent of 20?

4. What percent of 160 is 60?

5. What percent of 21 is 63?

6. In a certain class, 15 students answered all the questions correctly on a test. If there were 25 students in the class, what percent of the students answered all the questions correctly?

7. A car sells for $3250 cash. It can also be purchased for a down payment of $487.50 plus monthly installments. What percent is required for a down payment?

C.6 FIND A PERCENT OF A NUMBER

EXAMPLE 1

Find 15% of 60.

15% of 60 = ?

Solution:

In order to proceed we change the percent to either a fraction or a decimal and multiply. (In most cases it is easier to multiply by the decimal.) We will do it *both* ways here.

$$15\% = .15 = \frac{15}{100} = \frac{3}{20}$$

As a fraction:

$$\frac{3}{20} \times \frac{60}{1} = \frac{180}{20} = 9$$

$$15\% = \frac{3}{20}$$

As a decimal:

$$
\begin{array}{r}
60 \\
\times\ .15 \\
\hline
300 \\
60 \\
\hline
9.00
\end{array}
$$

$$15\% = .15$$

Therefore, 15% of 60 is 9.

EXAMPLE 2

32% of $423 = ?

What is 32% of $423?

Solution:

Change 32% to a decimal and multiply.

32% = .32

$$32\% = .32$$

$$
\begin{array}{r}
\$423 \\
\times\ .32 \\
\hline
846 \\
1269 \\
\hline
\$135.36
\end{array}
$$

Therefore, 32% of $423 is $135.36.

EXAMPLE 3

12% simple interest.

Phil borrows $2000 from a bank at a simple interest rate of 12% per year. How much *interest* does Phil owe at the end of the year?

Solution:

12% of $2000 = ?

The interest rate means that Phil has to pay 12% of $2000 each year in order to borrow the money. The amount of interest he owes after one year is 12% of $2000.

12% = .12

$$
\begin{array}{r}
\$2000 \\
\times\ .12 \\
\hline
4000 \\
2000 \\
\hline
\$240.00
\end{array}
$$

Therefore, Phil owes $240 in interest charges. (In addition to the $2000 he borrowed.)

EXERCISES C.6

1. What is 25% of 92?

2. What is 7% of 112?

3. What is 150% of 24?

4. What is $12\frac{1}{2}$% of 220?

5. What is 45% of 6.25?

6. Alice took a poll of the freshmen at Dover College. She found that of the 1160 freshmen, 70% liked cottage cheese. How many freshmen liked cottage cheese?

7. A store is advertising a washing machine on sale at a reduction of 20%. The machine originally sold for $248.

   a) How much money would you "save" if you bought during the sale?

   b) What is the selling price during the sale?

8. A real estate broker receives a commission of 6% if he sells a house. If the broker sells a house for $42,000, how much will the homeowner receive?

C.7 FIND A NUMBER WHEN A PERCENT OF IT IS KNOWN

EXAMPLE 1

25% of ? = 40.

25% of what number is 40?

Solution:

First, we must be certain that we understand what the problem is asking.

.25 × N = 40.

We are told that 25% of some number N is 40. This means that if we *multiplied* N by 25% (or .25) we would get 40 (as in Section C.6).

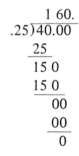

So, to find N we *reverse* the process and *divide* 40 by 25% (or .25).

$$
\begin{array}{r}
1\,60. \\
.25\overline{)40.00} \\
\underline{25} \\
15\,0 \\
\underline{15\,0} \\
00 \\
\underline{00} \\
0
\end{array}
$$

.25 × 160 = 40.

Therefore, 40 is 25% of 160.

Note:
To check the solution, determine 25% of 160.

$$
\begin{array}{r}
160 \\
\times\ .25 \\
\hline
800 \\
320 \\
\hline
40.00
\end{array}
$$

25% of 160 is 40 — so it checks.

EXAMPLE 2

63 is 38% of what number?

Solution:

This means that 38% of some number N is 63. That is, $.38 \times N = 63$.

Reverse the operation and divide 63 by 38%.

$$
\begin{array}{r}
1\ 65.78 \\
.38\overline{)63.00.00} \\
\underline{38} \\
25\ 0 \\
\underline{22\ 8} \\
2\ 20 \\
\underline{1\ 90} \\
30\ 0 \\
\underline{26\ 6} \\
3\ 40 \\
\underline{3\ 04} \\
36
\end{array}
$$

Therefore, 63 is 38% of 165.8 to the *nearest tenth*.

We could check by doing the following:

$$.38 \times 165.8$$

But remember that 165.8 is an *approximation* so the answer should be *approximately* 63. Try it.

38% of ? = 63

$.38 \times N = 63$

Reverse the operation.

$.38 \times 165.8 \approx 63.$

PASS
WITH
CARE

EXAMPLE 3

At the last election in Plainview 47,012 people voted. This number represented 73% of the eligible voters. How many eligible voters were there?

73% of ? = 47,012.

Solution:

This means that 73% of some number N is 47,012.

$$.73 \times N = 47,012.$$

.73 × N = 47,012.

Reverse the operation.

Reverse the operation and *divide* 47,012 by .73.

```
          644 00.
.73)47012.00.
    438
    321
    292
    292
    292
      0
```

.73 × 64,400 = 47,012.

Therefore, there were 64,400 eligible voters.

EXERCISES C.7

1. 32 is 80% of what number?
2. 15 is 6% of what number?
3. 120% of what number is 30?
4. $12\frac{1}{2}$% of what number is 42?
5. 8.6% of what number is 12.9?
6. A salesman gets a commission of 20% on what he sells. How much would he have to sell in order to earn $550?
7. A coat is on sale for $81. This represents a 10% reduction from the original price. What was the original price?

Answers

CHAPTER ONE ANSWERS

Exercises 1.2

1. a. Yes, order of listing the outcomes is immaterial.
   c. No, the outcomes must be separated by commas.
   e. No
   g. No

2. a. $S = \{b,g,y,r\}$, $n(S) = 4$
   c. $S = \{a,b,\ldots\ldots,y,z\}$, $n(S) = 26$
   e. $S = \{$ Jan., Feb., $\ldots\ldots$, Nov., Dec.$\}$, $n(S) = 12$
   g. $S = \{0,1,2,3,4,5,6,7,8,9\}$, $n(S) = 10$
   i. $S = \{g_1, g_2, \ldots\ldots, g_9, g_{10}\}$, $n(S) = 10$

3. a. $S = \{bg, by, br, gy, gr, yr\}$, $n(S) = 6$
   or $\{rr, rg, rb, gg, gb\}$, $n(S) = 5$
   c. $S = pn, pd, pq, nd, nq, dq\}$, $n(S) = 6$
   e. $S = g_1 g_2, g_1 g_3, \ldots\ldots, g_8 g_{10}, g_9 g_{10}\}$, $n(S) = 45$

Exercises 1.3

1. a) An outcome. A simple event is a set.
   c) Yes, It's the number of outcomes. You cannot have a fraction of an outcome.
   e) No. A set can only be a subset of another set if *every* outcome of the subset is in the other set.
   g) Neither.   has no outcomes.

2. a) The sum on the two dice is 3.
   c) The number on the second die is 4.
   e) The number on both dice is the same.

3. a) $A = \{H_1, H_2, H_3, H_4, H_5, H_6\}$
   c) $C = \{H_2, H_3, H_4, H_5, H_6, T_2, T_3, T_4, T_5, T_6\}$
   e) $E = \{H_3, T_3, H_4, T_4, H_5, T_5\}$

4.   Note that each of questions a,b,c,d,e,g admits more than one
     answer.
     a) $A = \{r, b, o, y\}$
     c) $A = \{r,b\}$, $B = \{o,y\}$
     e) $V = \{r,o,p\}$, $W = \{r,b\}$ W V.
     g) $A = \{r\}$
5. a) $S = \{g_1g_2b_1, g_1g_2b_2, g_1g_2b_3, g_1b_1b_2, g_1b_1b_3, g_1b_2b_3, g_2b_1b_2,$
       $g_2b_1b_3, g_2b_2b_3, b_1b_2b_3\}$, $n(S) = 10.$
   c) $\{g_1g_2b_2, g_1g_2b_2, g_1g_2b_3, g_2b_1b_2, g_2b_1b_3, g_2b_2b_3\}$
   e) $\{g_1g_2b_3\}$

Exercises 1.4

1. a) Yes. The probability of an event can be any value between 0 and
      1.
   c) Yes
   e) No $\dfrac{17}{16} > 1$

2. b) $\dfrac{2}{6} \approx 33\%$

3. b) $\dfrac{2}{12} = \dfrac{1}{6} \approx 17\%$

4. a) $\dfrac{1}{5} = 20\%$ c) $\dfrac{2}{5} = 40\%$ e) $\dfrac{4}{5} = 80\%$

5. a) $\dfrac{1}{36} \approx 2.7\% \approx 3\%$ c) $\dfrac{3}{36} \approx 8.3\% \approx 8\%$ e) $\dfrac{5}{36} \approx 13.8\% \approx 14\%$

   g) $\dfrac{5}{36} \approx 14\%$ i) $\dfrac{3}{36} \approx 8\%$ k) $\dfrac{1}{36} \approx 3\%$

6. b)   The event that the sum on the two dice is 2 and
         the event that the sum on the two dice is 12.
   c) Sum of 2 and sum of 12.
      Sum of 3 and sum of 11.
      Sum of 4 and sum of 10.
      Sum of 5 and sum of 9.
      Sum of 6 and sum of 8.

7. a) $\dfrac{1}{52} \approx 1.9\% \approx 2\%$ c) $\dfrac{13}{52} = \dfrac{1}{4} = 25\%$ e) $\dfrac{12}{52} \approx 23\%$

   g) $\dfrac{20}{52} = \dfrac{5}{13} \approx 38.4\% \approx 38\%$ i) $\dfrac{2}{13} \approx 15.3\% \approx 15\%$

Exercises 1.5

1. We may write the sample space as

$$S = \{w_1, w_2, r_1, r_2, r_3, g_1, g_2, g_3, g_4, b_1, b_2, b_3, b_4, b_5\}$$

a) $\dfrac{2}{12} = \dfrac{1}{7}$ c) $\dfrac{4}{14} = \dfrac{2}{7}$ e) $\dfrac{5}{14}$

g) $\dfrac{10}{14} = \dfrac{5}{7}$ i) $\dfrac{11}{14}$

3. $S = \{mon, moe, moy, mne, mny, mey, one, ony, oey, ney\}$, $n(S) = 10$

a) $\{mom, mny, one, ony, ney, mne\}$, $\dfrac{6}{10} = \dfrac{3}{5}$

5. a) $P(A) = \dfrac{4}{52} = \dfrac{1}{13}$ c) $P(C) = \dfrac{12}{52} = \dfrac{3}{13}$ e) $P(E) = \dfrac{39}{52} = \dfrac{3}{4}$

6. a) $A = \{(1,1),(1,3), (2,2), (3,1), (1,5), (2,4), (3,3), (4,2), (5,1),$
$(2,6), (3,5),(4,4),(5,3), (6,2), (4,6), (5,5), (6,4), (6,6)\}$

c) $C = \{(1,4), (2,3), (3,2), (4,1), (4,6), (5,5), (6,4)\}$

e) $E = \{(1,1), (1,3), (2,2), (3,1), (1,5), (2,4), (3,3), (4,2), (5,1),$
$(2,6), (3,5), (4,4), (5,3), (6,2), (4,6), (5,5), (6,4), (6,6),$
$(1,2), (2,1), (3,6), (4,5), (5,4), (6,3)\}$

7. a) $\dfrac{18}{36} = \dfrac{1}{2}$ c) $\dfrac{7}{36}$ e) $\dfrac{24}{36} = \dfrac{2}{3}$

(note that $\dfrac{24}{36} = \dfrac{18}{36} + \dfrac{12}{36} - \dfrac{6}{36}$

CHAPTER TWO ANSWERS

Exercises 2.2

1. $3 \times 4 = 12$
3. $5 \times 4 = 20$
5. $\dfrac{5 \times 4}{2} = \dfrac{20}{2} = 10$
7. $n(S) = 4 \times 4 = 16$

a) $\dfrac{2}{16} = \dfrac{1}{8}$

b) $\dfrac{1}{16}$

c) No. 2 ways for *(a)* = *gb + bg*

1 way for *(b)* = *bb*

8. a) $10 \times 10 = 100$

b) $\dfrac{10}{100}$

10. a) (i.) $10 \times 10 = 100$
      (ii.) $10 \times 9 = 90$

b) $\dfrac{90}{100} = \dfrac{9}{10}$

Exercises 2.3

1. $3 \times 3 \times 3 = 27$
3. $10 \times 15 \times 20 = 3000$
5. $5 \times 5 \times 5 \times 5 = 625$
7. $26 \times 10 \times 9 \times 8 = 18,720$ with no replacement
    $26 \times 9 \times$ ooops
9. a) $10 \times 9 \times 8 = 720$
  b) $10 \times 10 \times 10 = 1000$
  c) $9 \times 9 \times 8 = 648$

d) $\dfrac{720}{1000} = \dfrac{18}{25}$

11. a) $\dfrac{1}{8}$ b) $\dfrac{3}{8}$

Exercises 2.4

1.
3. a) $C = \{\, thh,\ hht,\ htt,\ tht,\ tth \,\}$
  b) C
5. $n(\overline{A}) = n(S) - n(A) = 30 - 12 = 18$

7. $P(\overline{A}) = 1 - P(A) = \dfrac{9}{13}$

9. a) $\dfrac{3}{4}$ b) $\dfrac{1}{4}$
11. $1 - .4 = .6$
13. a) $1 - .411 - .589 \approx 59\%$
   b) $1 - .960 = .040$ 4%

CHAPTER THREE ANSWERS

Exercises 3.2

1. $\frac{1}{13} \times 1 \approx \$ .08$

3. a) $\frac{1}{100} \times 75 = \$ .75$ No, he should pay $ .75

   b) $1 - .75 = \$ .25$

5. $\frac{1}{1000} \times 150 = \$ .15$ Yes, a ticket should cost $ .15

7. $.002 \times 10{,}000 = \$20$

9. a) No. $\frac{15}{36} \times 2 \ \ .83$ The player should pay $.83 for Over or Under

   $\frac{6}{36} \times 5 \ \ .83$ The play should pay $ .83 to play Seven

   b) None. He loses the same on each game.

Exercises 3.3

1. $\frac{1}{4} (10 + 5 + 1 + 0) = \$4$

3. $\frac{1}{1000} (100 + 50 + 50 + 50) = \$.25$

5. $\left(\frac{11}{36} \times 2 \right) + \frac{6}{36} \times 1 = \$ .78$

7. $\left(\frac{1}{8} \times 1.50 \right) + \frac{3}{8} \times 1) + \frac{3}{8} \times .50) = \$ .75$

9. $\left(\frac{1}{81} \times 5 \right) + \left(\frac{2 \times (1 \times 3)}{81} \times 3 \right) \ \$ .28$

11. a) $\dfrac{P(A)}{P(\overline{A})} = \dfrac{\frac{2}{4}}{\frac{2}{4}} = \dfrac{1}{1}$ b) $\dfrac{P(\overline{A})}{P(A)} = \dfrac{\frac{2}{4}}{\frac{2}{4}} = \dfrac{1}{1}$ c) $\dfrac{P(B)}{P(\overline{B})} = \dfrac{\frac{2}{4}}{\frac{2}{4}} = \dfrac{1}{1}$

   d) $\dfrac{P(\overline{B})}{P(B)} = \dfrac{\frac{2}{4}}{\frac{2}{4}} = \dfrac{1}{1}$

Exercises 3.4

1. a) $P(A) = \dfrac{4}{52}$

   $E = 100 \text{ x } \dfrac{4}{52} \approx 7.6 \approx 8$

   c) $P(C) = \dfrac{12}{52}$

   $E = 100 \text{ x } \dfrac{12}{52} \approx 23.0 = 23$

   e) $P(E) = \dfrac{39}{52}$

   $E = 100 \text{ x } \dfrac{39}{52} = 75$

2. a) $P(A) = \dfrac{6}{36}$

   $E = 60 \text{ x } \dfrac{6}{36} = 10$

   c) $P(C) = \dfrac{30}{36}$

   $E = 60 \text{ x } \dfrac{30}{36} = 50$

3. $P(D) = .02; \ E = 500 \text{ x } 0.02 = 10$
5. $P(U) = 0.002; \ E = 500 \text{ x } 1.002 = 1$

CHAPTER FOUR ANSWERS

Exercises 4.3

2. a) $\dfrac{310}{2163} \approx 14\%$ c) $\dfrac{580}{2163} \approx 27\%$ e) $\dfrac{823}{2163} \approx 38\%$

4. a) $\dfrac{58}{116} = \dfrac{1}{2}$ c) $\dfrac{23}{116}$

Exercises 4.4

1. a)   .48 (mean relative frequency)
   b)   .52 (mean relative frequency)
   c)   Yes, since .48 and .52 are quite close to each other.
2. a) 175, 157, 149, 140, 139, 135, 130, 100, 98 (or in ascending order)

mean = 136, median = 139

c) mean = $\dfrac{71}{100}$, median = $\dfrac{3}{4}$

4.   mean   21.6 mpg

Exercises 4.5

2. $\dfrac{623}{1000}$, $\dfrac{377}{1000}$

3. a) 1 - .00179 = .99821

4. a) $\dfrac{1241}{2163}$ c) $\dfrac{230}{2163}$ e) $\dfrac{251}{1241}$, $n(S) = 1241$

ANSWERS TO APPENDICES

Exercises A.1

1. a) $\dfrac{1}{3}$ c) $\dfrac{2}{5}$

Exercises A.2

1. $\dfrac{2}{6}$, $\dfrac{3}{9}$, etc.   3. $\dfrac{6}{16}$, $\dfrac{9}{24}$, etc.   5. $\dfrac{3}{21}$ 7. $\dfrac{9}{15}$ 9. $\dfrac{20}{24}$

11. $\dfrac{24}{42}$

Exercises A.3

1. $\dfrac{1}{2} < \dfrac{3}{4}$ 3. $\dfrac{5}{9} < \dfrac{7}{9}$ 5. $\dfrac{2}{3} = \dfrac{6}{9}$

7. $\dfrac{7}{8} < \dfrac{9}{10}$ 11. $\dfrac{2}{9} < \dfrac{4}{7} < \dfrac{3}{5} < \dfrac{2}{3}$

Exercises A.4

1. $\dfrac{1}{3}$ 3. $\dfrac{3}{4}$ 5. $\dfrac{6}{13}$ 7. $\dfrac{3}{5}$ 9. $\dfrac{2}{3}$ 11. a) $\dfrac{20}{25} = \dfrac{4}{5}$ b) $\dfrac{1}{5}$

Exercises A.5

1. $\dfrac{7}{10}$ 3. $\dfrac{29}{35}$ 5. $\dfrac{13}{28}$ 7. $\dfrac{11}{9}$ 9. $\dfrac{7}{30}$ 11. $\dfrac{47}{36}$ 13. $\dfrac{2}{21}$

Exercises A.6

1. $\dfrac{6}{7}$ 3. $\dfrac{6}{35}$ 5. $\dfrac{4}{7}$ 7. $\dfrac{5}{63}$ 9. $\dfrac{1}{24}$ 11. $1.05

Exercises A.7

1. $\dfrac{5}{18}$ 3. $\dfrac{1}{9}$ 5. $\dfrac{7}{8}$ 7. $\dfrac{7}{22}$ 9. $\dfrac{1}{14}$ 11. 6

Exercoses A.8

1. $\dfrac{7}{2}$ 3. $\dfrac{39}{4}$ 5. $\dfrac{130}{9}$ 7. $\dfrac{377}{6}$ 9. $4\dfrac{1}{2}$ 11. $4\dfrac{4}{7}$ 13. $6\dfrac{3}{4}$ 15. 7

Exercises A.9

1. $5\dfrac{7}{15}$ 3. $22\dfrac{1}{28}$ 5. $4\dfrac{1}{6}$ 7. $6\dfrac{1}{14}$ 9. $\dfrac{27}{8} = 3\dfrac{3}{8}$ 11. $\dfrac{1298}{9} = 144\dfrac{2}{9}$

13. $\dfrac{1}{4}$ 15. $\dfrac{10}{3} = 3\dfrac{1}{3}$ 17. $6\dfrac{1}{12}$ 19. $\dfrac{629}{8} = 78\dfrac{5}{8}$

Exercises B.1

1. six tenths
3. eight hundredths
5. nine and forty one hundredths
7. seventy six and twelve hundredths
9. one hundred eight and five tenths

Exercises B.2

1. $\dfrac{4}{5}$ 3. $\dfrac{7}{8}$ 5. $\dfrac{57}{100}$ 7. $23\dfrac{3}{4}$ 9. $\dfrac{9}{2000}$

Exercises B.3

1.   .5   3. .75   5. .6   7. .625 9. .05

Exercises B.4

1. 6.24, 6.2, 6   3. 79.61, 79.6, 80
5. .07, .1, 0   7. 5.35, 5.3, 3

Exercises B.5

1. 138.8   3. 289.692   5. 51.751   7. 5.804   9. 235.37
11. 40.1 seconds

Exercises B.6

1. 3892.4   3. 1.48995   5. 1.72244   7. 1.1808
9. 105.2724 miles

Exercises B.7

1. 15.61   3. 25.6   5. 2.76   7. 1609.36   9. 37.5 gallons

Exercises C.1

1.   23% means "23 per hundred".   So she sold 23
      bars of candy.
3.   He owns 300 acres, so he plows "20 acres per hundred" since
      20 x 3 = 60. So he plowed 20% of his land.

Exercises C.2

1. 37%   2. 54%   3. 40%   5. 3.8% or $3\frac{4}{5}$%   7. 400%
9. 750%   11. 40%   13. 67%   15. 22%   17. 42%
19. 64%

Exercises C.3

1. $\frac{3}{10}$ 3. $\frac{2}{25}$ 5. $\frac{1}{8}$ 7. $\frac{1}{125}$ 9. $\frac{18}{25}$ 11. .30   13. .08
15. .125   17. .008   19. .75

Exercises C.5

1. 25%   3. 125% 5. 300%   6. 60%

Exercises C.6

1. 23.5   3. 36   5. 2.8125   7. a) $49.60
                                       b) $198.40

Exercises C.7

1. 40   3. 25   5. 150   6. $2750

Index